I0488447

Evaluation of Quality-Control Data Collected by the U.S. Geological Survey for Routine Water-Quality Activities at the Idaho National Laboratory, Idaho, 1996–2001

By Gordon W. Rattray

DOE/ID-22222
Prepared in cooperation with the U.S. Department of Energy

Scientific Investigations Report 2012–5270

U.S. Department of the Interior
U.S. Geological Survey

U.S. Department of the Interior
KEN SALAZAR, Secretary

U.S. Geological Survey
Marcia K. McNutt, Director

U.S. Geological Survey, Reston, Virginia: 2012

For more information on the USGS—the Federal source for science about the Earth, its natural and living resources, natural hazards, and the environment, visit http://www.usgs.gov or call 1–888–ASK–USGS.

For an overview of USGS information products, including maps, imagery, and publications, visit http://www.usgs.gov/pubprod

To order this and other USGS information products, visit http://store.usgs.gov

Suggested citation:
Rattray, G.W., 2012, Evaluation of quality-control data collected by the U.S. Geological Survey for routine water-quality activities at the Idaho National Laboratory, Idaho, 1996–2001: U.S. Geological Survey Scientific Investigations Report 2012-5270 (DOE/ID-22222), 74 p.

Contents

Figures

Tables

Tables—Continued

Conversion Factors, Datum, and Abbreviations and Acronyms

Conversion Factors

Inch/Pound to SI

Multiply	By	To obtain
Length		
mile (mi)	1.609	kilometer (km)
Area		
square mile (mi^2)	259.0	hectare (ha)
square mile (mi^2)	2.590	square kilometer (km^2)
Radioactivity		
picocurie per liter (pCi/L)	0.037	Becquerel per liter (Bq/L)

Concentrations of chemical constituents in water are given in picocuries per liter (pCi/L), milligrams per liter (mg/L), or micrograms per liter (μg/L).

Datum

Horizontal coordinate information is referenced to North American Datum of 1927 (NAD 27).

Conversion Factors, Datum, and Abbreviations and Acronyms—Continued

Abbreviations and Acronyms

Abbreviation or Acronym	Definition
ATRC	Advanced Test Reactor Complex
BC	Birch Creek
BLR	Big Lost River
CFA	Central Facilities Area
cl	confidence level
Cs-137	cesium-137
CSU	combined standard uncertainty
CV	coefficient of variation
DIW	deionized water
DOE	U.S. Department of Energy
ESRP	eastern Snake River Plain
IBW	inorganic blank water
INL	Idaho National Laboratory
INLPO	USGS Idaho National Laboratory Project Office
INTEC	Idaho Nuclear Technology and Engineering Center
LLR	Little Lost River
LRL	laboratory reporting level
LT-MDL	long-term method detection level
MDC	minimum detectable concentration
ML	Mud Lake
MFC	Materials and Fuels Complex
MRL	minimum reporting level
N	nitrogen
NAD	normalized absolute difference
NWQL	U.S. Geological Survey National Water Quality Laboratory
P	phosphorus
Pu-239	Plutonium-239
QA	quality assurance
QC	quality control
RESL	DOE Radiological and Environmental Sciences Laboratory
RSD	relative standard deviation
RWMC	Radioactive Water Managmenet Complex
Sr-90	strontium-90

Evaluation of Quality-Control Data Collected by the U.S. Geological Survey for Routine Water-Quality Activities at the Idaho National Laboratory, Idaho, 1996–2001

By Gordon W. Rattray

Abstract

The U.S. Geological Survey, in cooperation with the U.S. Department of Energy, collects surface water and groundwater samples at and near the Idaho National Laboratory as part of a routine, site-wide, water-quality monitoring program. Quality-control samples are collected as part of the program to ensure and document the quality of environmental data. From 1996 to 2001, quality-control samples consisting of 204 replicates and 27 blanks were collected at sampling sites. Paired measurements from replicates were used to calculate variability (as reproducibility and reliability) from sample collection and analysis of radiochemical, chemical, and organic constituents. Measurements from field and equipment blanks were used to estimate the potential contamination bias of constituents.

The reproducibility of measurements of constituents was calculated from paired measurements as the normalized absolute difference (NAD) or the relative standard deviation (RSD). The NADs and RSDs, as well as paired measurements with censored or estimated concentrations for which NADs and RSDs were not calculated, were compared to specified criteria to determine if the paired measurements had acceptable reproducibility. If the percentage of paired measurements with acceptable reproducibility for a constituent was greater than or equal to 90 percent, then the reproducibility for that constituent was considered acceptable for the period 1996–2001. The percentage of paired measurements with acceptable reproducibility was greater than or equal to 90 percent for all constituents except orthophosphate (89 percent), zinc (80 percent), hexavalent chromium (53 percent), and total organic carbon (TOC; 38 percent). The low reproducibility for orthophosphate and zinc was attributed to calculation of RSDs for replicates with low concentrations of these constituents. The low reproducibility for hexavalent chromium and TOC was attributed to the inability to preserve hexavalent chromium in water samples and high variability with the analytical method for TOC.

The reliability of measurements of constituents was estimated from pooled RSDs that were calculated for discrete concentration ranges for each constituent. Pooled RSDs of 15 to 33 percent were calculated for low concentrations of gross-beta radioactivity, strontium-90, ammonia, nitrite, orthophosphate, nickel, selenium, zinc, tetrachloroethene, and toluene. Lower pooled RSDs of 0 to 12 percent were calculated for all other concentration ranges of these constituents, and for all other constituents, except for one concentration range for gross-beta radioactivity, chloride, and nitrate + nitrite; two concentration ranges for hexavalent chromium; and TOC. Pooled RSDs for the 50 to 60 picocuries per liter concentration range of gross-beta radioactivity (reported as cesium-137) and the 10 to 60 milligrams per liter (mg/L) concentration range of nitrate + nitrite (reported as nitrogen [N]) were 17 percent. Chloride had a pooled RSD of 14 percent for the 20 to less than 60 mg/L concentration range. High pooled RSDs of 40 and 51 percent were calculated for two concentration ranges for hexavalent chromium and of 60 percent for TOC.

Measurements from (1) field blanks were used to estimate the potential bias associated with environmental samples from sample collection and analysis, (2) equipment blanks were used to estimate the potential bias from cross contamination of samples collected from wells where portable sampling equipment was used, and (3) a source-solution blank was used to verify that the deionized water source-solution was free of the constituents of interest. If more than one measurement was available, the bias was estimated using order statistics and the binomial probability distribution. The source-solution blank had a detectable concentration of hexavalent chromium of 2 micrograms per liter. If this bias was from a source other than the source solution, then about 84 percent of the 117 hexavalent chromium measurements from environmental samples could have a bias of 10 percent or more. Of the 14 field blanks that were collected, only chloride (0.2 milligrams per liter) and ammonia (0.03 milligrams per liter as nitrogen), in one blank each, had detectable concentrations. With an estimated confidence level of 95 percent, at least 80 percent of the 1,987 chloride concentrations measured from all environmental samples had a potential bias of less than 8 percent. The ammonia bias, which may have occurred at the analytical laboratory, could produce a potential bias of 5–150 percent in eight potentially affected ammonia measurements. Of the 12 equipment blanks that were collected, chloride was detected in 4 of these blanks, sodium in 3 blanks, and sulfate and hexavalent chromium were each detected in 1 blank. The concentration of hexavalent chromium in the equipment blank was the same concentration as in the source-solution blank collected on the same day, which indicates that the hexavalent chromium in the equipment blank is probably from a source other than the portable sampling equipment, such as the sample bottles or the source-solution water itself. The potential bias for chloride, sodium, and sulfate measurements was estimated for environmental samples that were collected using portable sampling equipment. For chloride, it was estimated with 93 percent confidence that at least 80 percent of the measurements had a bias of less than 18 percent. For sodium and sulfate, it was estimated with 91 percent confidence that at least 70 percent of the measurements had a bias of less than 12 and 5 percent, respectively.

Introduction

The Idaho National Laboratory (INL) was established by the U.S. Atomic Energy Commission—which later became the U.S. Department of Energy (DOE)—in 1949 for the development of atomic-energy applications, nuclear safety research, defense programs, and advanced energy concepts (Knobel and others, 2005, p. 1). The INL extends over approximately 890 mi^2 of the north-central part of the eastern Snake River Plain (ESRP) in southeastern Idaho (fig. 1) and overlies about 8 percent of the ESRP aquifer, which is a fractured basalt sole-source aquifer of significant economic value to the State of Idaho. During its operations, the INL has produced and discharged radiochemical and chemical wastes from site facilities to the unsaturated zone and the underlying aquifer through infiltration ponds, evaporation ponds and ditches, drain fields, injection wells, and burial sites (Bartholomay and Twining, 2010, p. 1).

The U.S. Geological Survey (USGS) began studying the water quality of the ESRP aquifer in 1949 as part of a program to characterize the water resources at the INL (Nace and others, 1959; Olmstead, 1962; Robertson and others, 1974). Sampling for radiochemical and chemical constituents was sporadic until 1964, when a water-quality monitoring network was established and routine (quarterly, semiannual, or annual) sample collection and analysis began (Knobel and others, 2005, p. 11). The monitoring network includes three separate monitoring programs: a routine, site-wide, water-quality monitoring program at and near the INL (figs. 2–4); a local monitoring program at the Naval Reactors Facility; and

off-site monitoring programs. The objectives of the water-quality monitoring network, which included both the aquifer and the perched groundwater zones, were to (1) monitor the concentrations and delineate the movement of facility-related radiochemical and chemical wastes, (2) understand the processes controlling the movement of the wastes, and (3) understand the processes controlling the groundwater chemistry (Mann, 1996, p. 2; Knobel and others, 2005, p. 1, 15, 20).

Beginning in 1980, field quality-control (QC) samples were routinely collected at groundwater and surface-water sites to ensure and document the quality of the environmental data. The collection of QC samples preceded documentation of a quality-assurance (QA) plan for water-quality activities by the USGS Idaho National Laboratory Project Office (INLPO) in 1989 (L.J. Mann, U.S. Geological Survey, written commun., 1989) and publication of the QA plan in 1996 (Mann, 1996). The QA plan described the collection of QC samples, such as replicates and blanks. Quality-control samples are an essential component of a water-quality monitoring program because data from QC samples can be used to identify, quantify, and document potential variability and bias, two types of errors in environmental data. The variability and bias "associated with environmental data must be known for the data to be interpreted properly and be scientifically defensible" (U.S. Geological Survey, 2006, p. 133).

Purpose and Scope

The purpose of this report is to investigate, and document, the quality of environmental data collected by the INLPO from 1996 to 2001. About 2,000 environmental, 204 field replicate, and 27 field blank samples were collected during this period. The quality of the environmental data was investigated by evaluating the replicate and blank data. Statistical analysis of constituent concentrations from replicates was used to calculate variability (as reproducibility and reliability) from sample collection and analysis of radiochemical, inorganic, and organic constituents. Similarly, statistical analysis of constituent concentrations from field and equipment blanks was used to estimate the potential bias from (1) sample collection and analysis of environmental samples and (2) cross contamination of environmental samples collected with portable sampling equipment.

Quality-control samples were collected independently for each of the monitoring programs, and QC data presented in this report were from samples collected for the routine, site-wide, water-quality monitoring program. However, the QC results and interpretations presented in this report also are applicable to chemical (but not radiochemical) data collected and analyzed for the off-site monitoring programs because these programs used the same field procedures, analytical methods, and laboratories for collecting and analyzing the chemical data as the routine monitoring program.

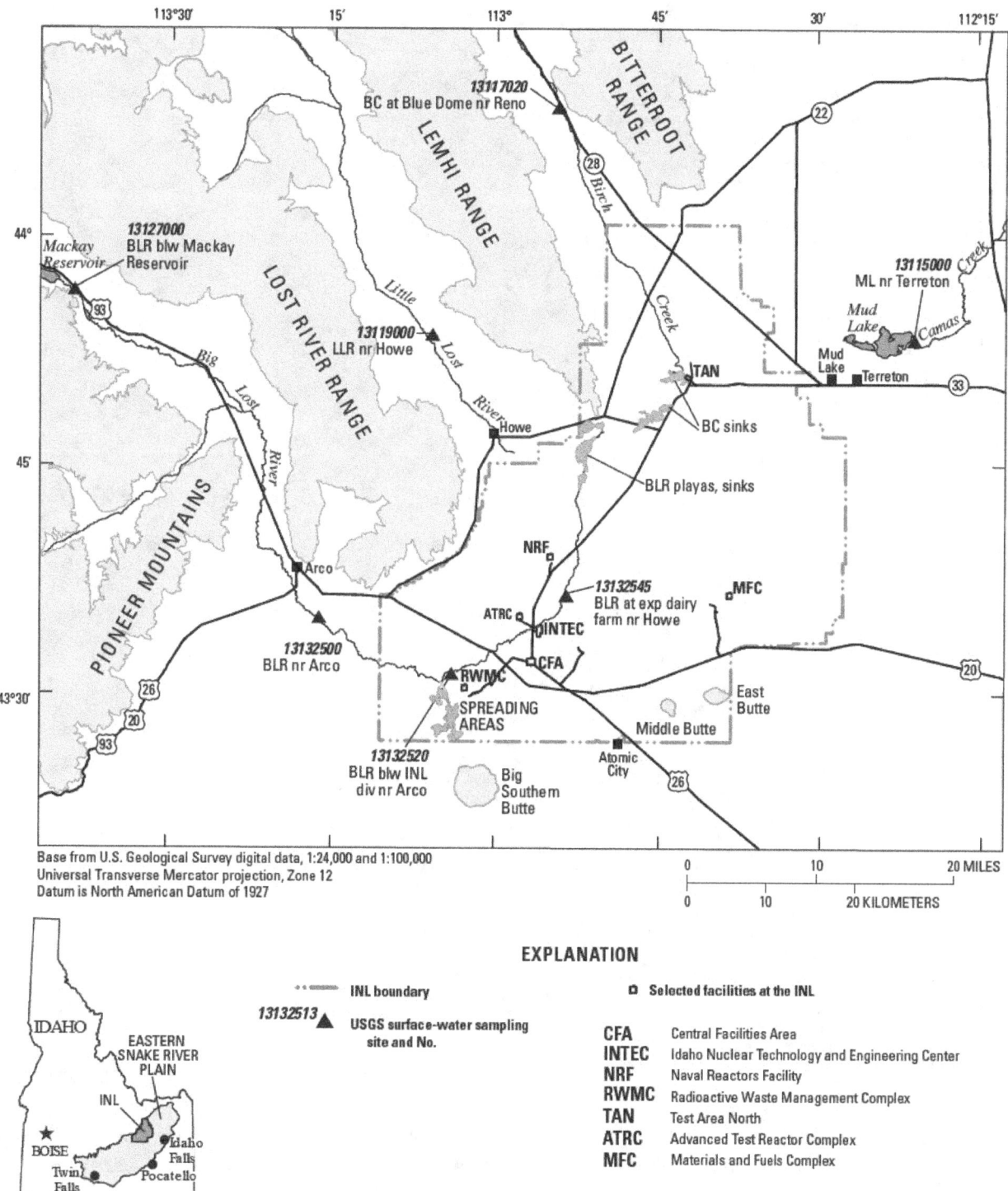

Figure 1.   Location of selected facilities and surface water sampling sites at the Idaho National Laboratory (INL) and vicinity, Idaho. Site names and abbreviations are listed in table 2.

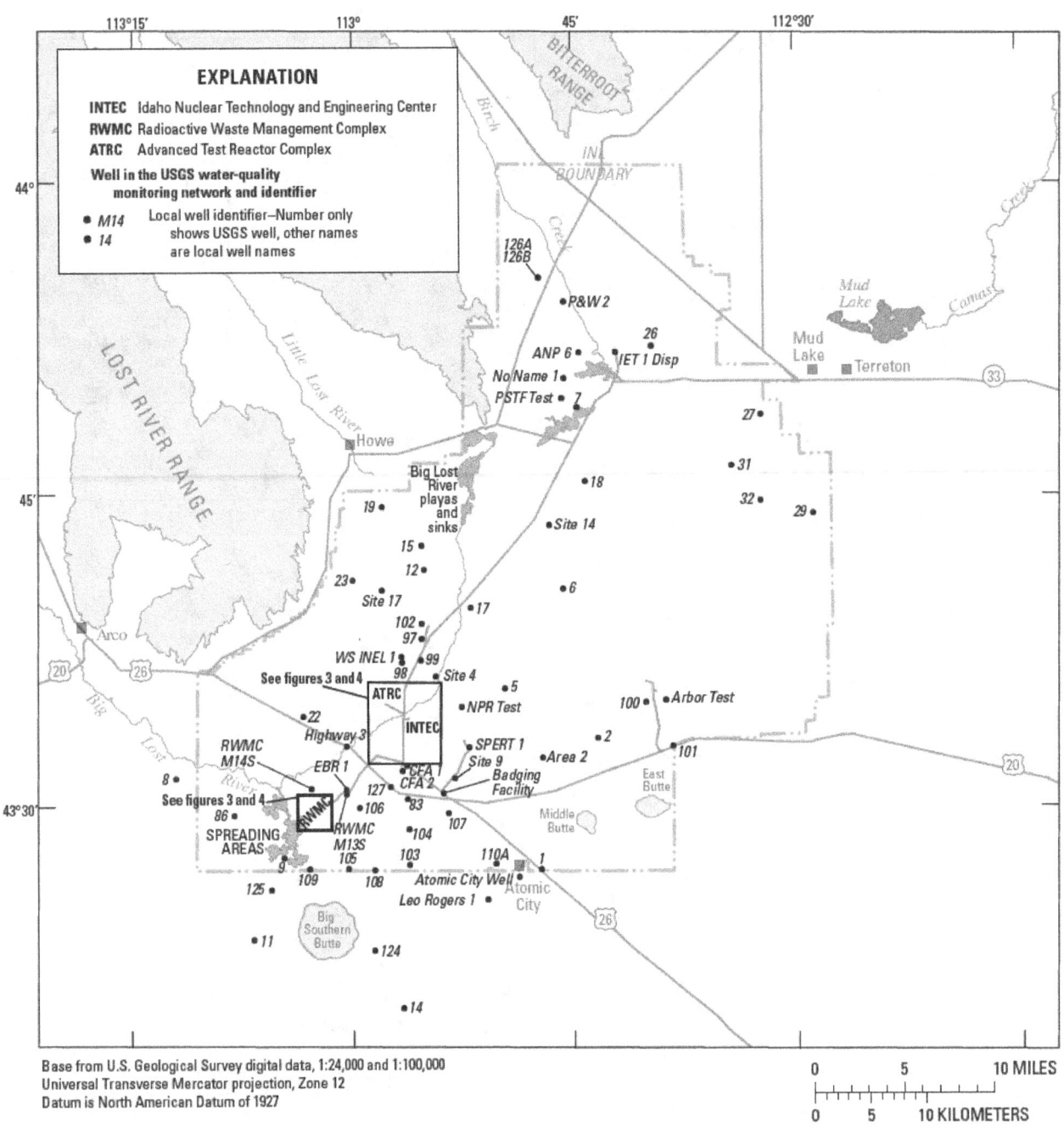

Figure 2. Location of aquifer wells in the U.S. Geological Survey (USGS) routine, site-wide, water-quality monitoring program at the Idaho National Laboratory (INL) and vicinity, Idaho, December 2001.

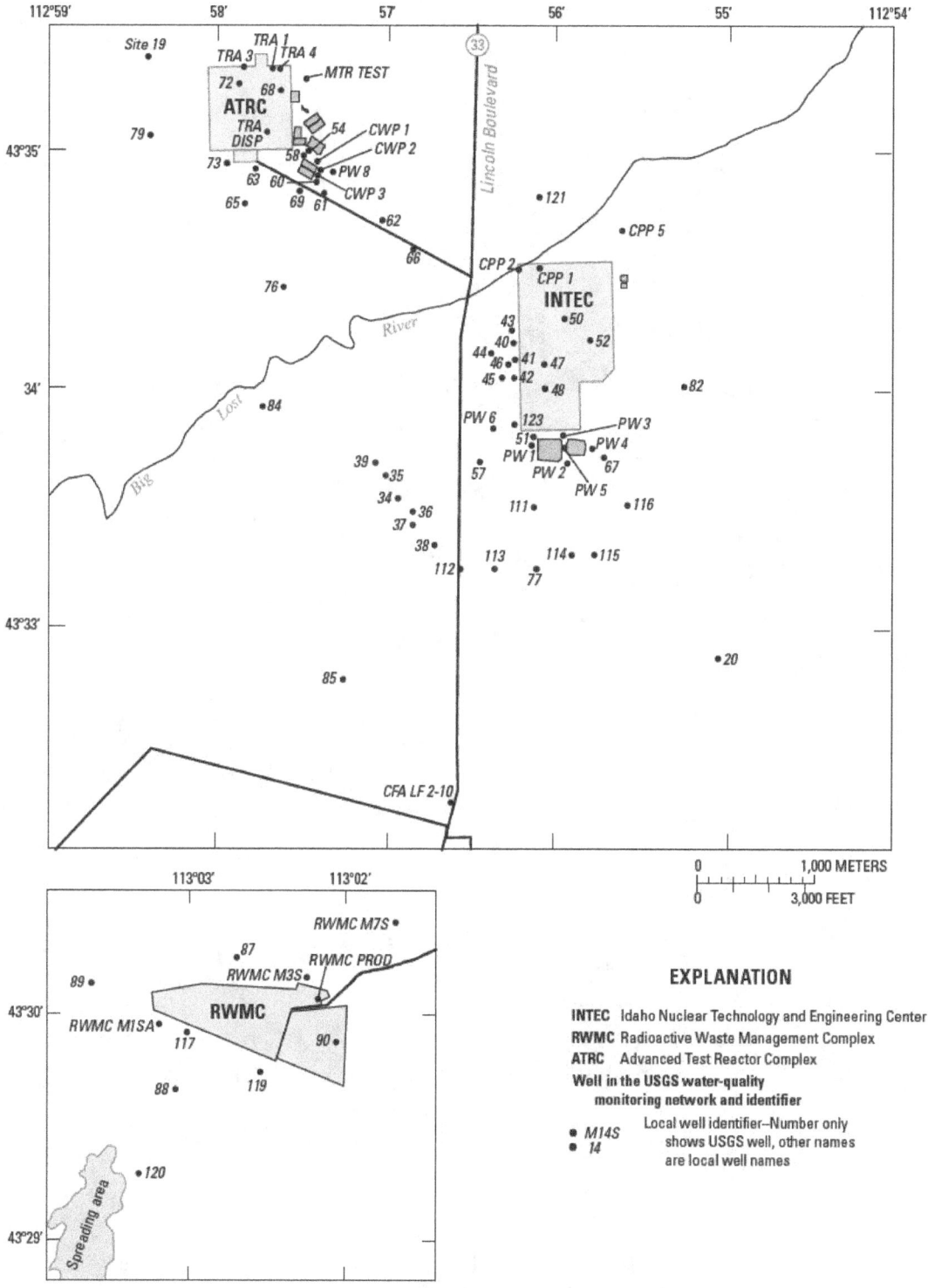

Figure 3. Location of aquifer wells in the U.S. Geological Survey (USGS) routine, site-wide, water-quality monitoring program at the Advanced Test Reactor Complex, Idaho Nuclear Technology and Engineering Center, and Radioactive Waste Management Complex at the Idaho National Laboratory, Idaho, December 2001.

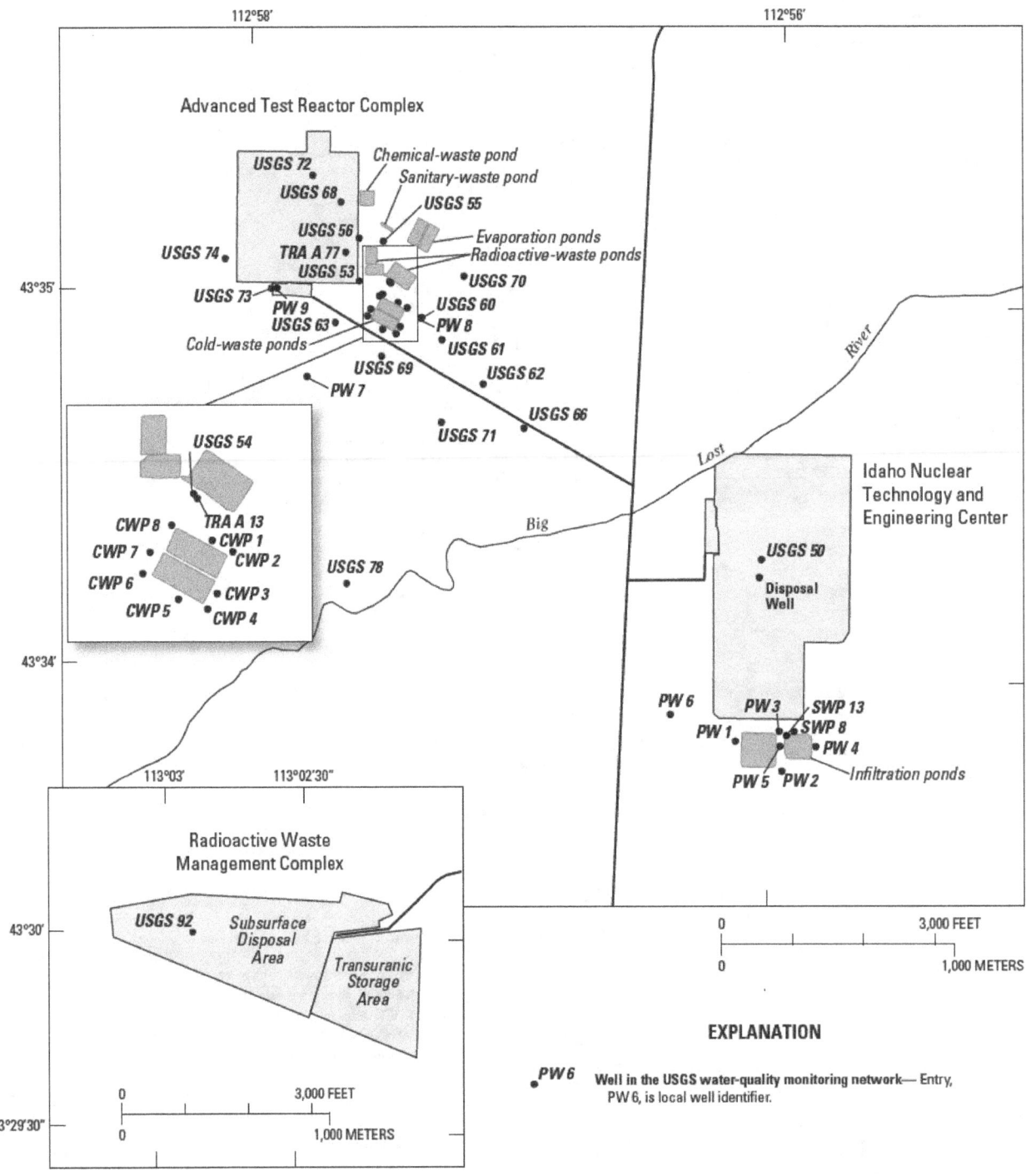

Figure 4.   Location of perched groundwater wells in the U.S. Geological Survey (USGS) routine, site-wide, water-quality monitoring program at the Idaho National Laboratory, Idaho, December 2001.

Previous Investigations

Quality-control results previously were presented in reports for the routine, on-site, monitoring program by Wegner (1989) and Williams (1996, 1997); for the local, on-site, monitoring program at the Naval Reactors Facility by Williams (1996, 1997) and Knobel and others (1999a); for off-site monitoring programs by Williams and others (1998) and Rattray and Campbell (2003); and for special on-site studies by Knobel and others (1999b) and Bartholomay and Twining (2010). Many of the results from previous reports are not directly comparable to the results presented in this report because the sample collection methods were different (Bartholomay and Twining, 2010) or laboratories other than the NWQL or the RESL were used for chemical or radiochemical analyses, respectively (Wegner, 1989; Williams, 1996; Williams and others, 1998; Knobel and others, 1999a, 1999b; Rattray and Campbell, 2003).

Methods

Collection of Quality-Control Samples

About 10 percent of the samples collected by the INLPO are dedicated to field QC samples (Mann, 1996, p. 12). Field QC samples collected from 1996 to 2001 included replicates and blanks.

Replicates

The INLPO generally collects a replicate from a sampling site with the longest interval of time, relative to all other sampling sites, since a replicate was last collected. This approximates a rotational sequence for collecting replicates from sampling sites and ensures that, during a period of several years, replicates are collected from most of the sampling sites. From 1996 to 2001, replicates were collected at 151 of the 173 sampling sites.

Replicates (also called replicate pairs) consisted of two water samples, an environmental sample and a replicate sample, with the replicate sample collected immediately after collection of the environmental sample. Replicates were submitted blind to the analytical laboratories, ensuring that the laboratories did not know the source of the water or that the samples were replicates. The replicate (and environmental) samples were collected in accordance with established sample-collection procedures and guidelines documented by the U.S. Geological Survey (2006) and the INLPO

quality-assurance plan (Mann, 1996). Equipment used during sample collection included sample bottles, capsule filters, Tygon® tubing, stainless steel pipes at wells with dedicated submersible pumps, bailers or a portable pump and tubing at wells without dedicated submersible pumps, and a churn splitter at surface water sites. Equipment and bottles were cleaned and rinsed in accordance with procedures described in Mann (1996, p. 6). Samples from pumped water were collected at wells after purging at least three wellbore volumes of water and after stable values of pH, specific conductivity, and temperature were measured. After sample collection, preservatives were added to sample bottles (if required), and the bottles were capped, labeled, chilled (if required), and stored in the USGS laboratory at the INL until delivery to the analytical laboratory. Sample bottles, with chain-of-custody forms, were mailed twice a week to the NWQL in sealed coolers and delivered by hand to the RESL at the end of each sampling event.

Blanks

The types of blanks collected included equipment, field, and source-solution blanks. The source solution blank was collected at the USGS laboratory at the INL. Equipment and field blanks were collected inside a USGS field camper at a sampling site.

Collection of equipment blanks consisted of field rinsing the bailer or portable pump and tubing in the same manner as for environmental samples—by rinsing the bailer with source solution or by passing a conditioning volume of source solution through the portable pump and tubing. For equipment blanks collected using the bailer, source solution was poured into the rinsed bailer and then poured from the bailer into a pitcher. For equipment blanks collected using the portable pump, source solution was pumped through the pump and tubing into a pitcher. For field and source-solution blanks, deionized water (DIW) source solution was poured into the pitcher for samples requiring filtration. The source solution was not filtered if inorganic blank water (IBW) was the source solution. Sample bottles for equipment, field, and source-solution blanks were then filled by pouring the source solution from the pitcher or source-solution bottle directly into the sample bottles, or by filtering DIW source solution into sample bottles. Bottles for all blank samples were preserved, stored, handled, and shipped as described for the replicates. Source solutions for the blanks consisted of DIW, which had been previously determined to be a suitable source solution for blanks (Williams, 1997, p. 23), and IBW purchased from the USGS Ocala Water Quality and Research Laboratory and certified to be free of the constituents of interest.

Analytical Methods and Data Reporting Conventions

The QC (and environmental) samples were submitted to the DOE Radiological and Environmental Services Laboratory (RESL) for analysis of radiochemical constituents and the USGS National Water Quality Laboratory (NWQL) for analysis of inorganic and organic constituents. Constituent analyses included:

- Radiochemical constituents

  - gross-alpha, gross-beta, and gamma radioactivity

  - tritium and strontium-90

  - plutonium-238, plutonium-239+240, and americium-241

- Inorganic constituents

  - major ions (sodium, chloride, fluoride, sulfate)

  - nutrients (ammonia, nitrate + nitrite, nitrite, orthophosphate)

  - metals (aluminum, antimony, arsenic, barium, beryllium, cadmium, cobalt, copper, lead, manganese, mercury, molybdenum, nickel, selenium, silver, thallium, uranium, zinc, total dissolved chromium, hexavalent chromium)

- Organic constituents

  - volatile organic compounds (VOCs)

  - total organic carbon (TOC).

Analytical methods used by the RESL are described by Bodnar and Percival (1982) and U.S. Department of Energy (1995). Analytical methods used by the NWQL are described by Goerlitz and Brown (1972), Thatcher and others (1977), Skougstad and others (1979), Barnett and Mallory (1971), Wershaw and others (1987), Fishman and Friedman (1989), Faires (1992), Fishman (1993), and Rose and Schroeder (1995).

Laboratory QA/QC practices are described in analytical method documents as well as by Bodnar and Percival (1982) and the U.S. Department of Energy (1995) for the RESL and Friedman and Erdmann (1982) and Pritt and Raese (1995) for the NWQL. Summaries of NWQL QC data for 1996–2001 are presented by Ludtke and others (1999, 2000) and the U.S. Geological Survey (2012a, b).

The RESL reports combined standard uncertainties (CSUs) with their analytical results. These propagated random uncertainties were calculated using variables such as yields, appropriate half-lives, counting efficiencies, and count times and are reported at a confidence level of one standard deviation (Williams, 1997, p. 10). A lower CSU relative to the result indicates a lower measurement uncertainty, and a higher CSU relative to the result indicates a higher measurement uncertainty.

Reporting levels used by the NWQL include minimum reporting levels (MRLs), long-term method detection levels (LT-MDLs), and laboratory reporting levels (LRLs) (Childress and others, 1999). The MRL is the smallest measured constituent concentration that can be reliably reported using a specific analytical method (Timme, 1995). The LT-MDL is determined by calculating the standard deviation of a sample with at least 24 spike sample measurements over an extended period of time (Childress and others, 1999, p. 19). The LRL generally is equal to twice the yearly-determined LT-MDL (Childress and others, 1999, p. 19). Results that are between the LT-MDL and the LRL, or between the LRL and the lowest calibration standard, are reported with the "E" remark code (Childress and others, 1999, p. 9), which means the result is estimated and has a greater uncertainty than data without the "E" remark. Non-detections were reported by the NWQL as censored values (reported with the "<" symbol) that were less than the MRL or LRL. Table 1 (at back of report) lists the MRLs and LRLs for the inorganic and organic constituents discussed in this report.

Statistical Methods

The normalized absolute difference (NAD) and the relative standard deviation (RSD) were calculated to estimate the reproducibility of radiochemical and chemical measurements, respectively. Pooled RSDs were calculated to estimate the reliability of measurements for all constituents. The potential bias of environmental samples was estimated from constituent concentrations from blank samples using order statistics and binomial probability.

Normalized Absolute Difference

Normalized absolute differences were calculated from radiochemical concentrations and their CSUs. The NAD then was used to test the null hypothesis that a pair of radiochemical measurements did not differ significantly when compared to their CSUs (Williams, 1996, p. 11–15; Parr and Porterfield, 1997, p. 30; McCurdy and others, 2008, p. 15). Instead of setting a value approximately equal to two times the CSU as a test of equivalence, the significance level, which indicates the weight of the evidence to accept or reject the null hypothesis of $x \pm CSU_x = y \pm CSU_y$, was determined using the NAD as the test statistic. At an NAD of 1.96, the significance level was 0.05 (assuming a normal distribution and a two-tailed test), the probability of error was 0.05, and the decision of whether or not concentrations were the same was determined at the 95-percent confidence level. Thus, for an NAD less than or equal to 1.96, the NAD was within the 95-percent confidence interval, the null hypothesis was accepted, and the concentrations did not differ significantly. Concentrations were considered significantly different when the NAD was greater than 1.96.

The equation for calculating the NAD is:

$$NAD = \frac{|x - y|}{\sqrt{CSU_x^2 + CSU_y^2}} \qquad (1)$$

where

x		is the concentration of a radiochemical in the environmental sample,
y		is the concentration of the same radiochemical in the replicate sample,
CSU_x		is the combined standard uncertainty of x at the 1σ confidence level, and
CSU_y		is the combined standard uncertainty of y at the 1σ confidence level.

Relative Standard Deviation

The RSD is the percent coefficient of variation (CV) and was calculated as (Taylor, 1987):

$$RSD = CV \times 100 \text{ percent} \qquad (2)$$

The CV was calculated as:

$$CV = \frac{s}{\overline{x}} \qquad (3)$$

where

s		is the standard deviation for a constituent from a replicate pair, and
$\overline{x}$		is the mean concentration for the same constituent and replicate pair.

The standard deviations and mean concentrations for constituents from replicate pairs were calculated as:

$$s = \sqrt{\frac{\sum_{i=1}^{n}(x_i - \overline{x})^2}{n-1}} \qquad (4)$$

and

$$\overline{x} = \frac{\sum_{i=1}^{n}x_i}{n} \qquad (5)$$

where

x_i		is a constituent concentration from the replicate pair, and
n		is 2, the number of constituent concentrations from the replicate pair.

The standard deviation and mean concentration used for calculating pooled relative standard deviations were calculated as:

$$s_{pooled} = \sqrt{\frac{\sum_{i=1}^{k}v_i s_i^2}{\sum_{i=1}^{k}v_i}} \qquad (6)$$

and

$$\overline{x}_{pooled} \sum_{i=1}^{2k}\frac{x_i}{2k} \qquad (7)$$

where

k		is the number of replicate pairs with results for the constituent of interest, and
v		is the degrees of freedom for s_{pooled} and is equal to k.

Binomial Probability

The distribution of constituent concentrations from blank samples was highly skewed, so a non-parametric statistical method was used to estimate the potential bias of constituents from blank sample measurements. The statistical method used here, using order statistics (with the ranking from low to high concentration) and binomial probability (Mueller, 1998, p. 5–6), determined a one-sided confidence interval, or a confidence level (cl) that represented "the probability that m observed values from a total of n observations are less than or equal to the $100p$th percentile of the sampled population" (Mueller, 1998, p. 5). The confidence level was calculated as:

$$cl = Prob\ (n, m, p) \qquad (8)$$

At the $100cl$, the concentration of the $m+1$ ranked observation represented the concentration that exceeded $100p$ percent of the values in the population. Because of the small number of blank samples collected, p-values of 0.70 and 0.80 were used instead of the more inclusive value of 0.95 and the $m+1$ ranked observation was always equal to the nth ranked observation. For example, for a set of 13 field blanks ($n = 13$), any contamination bias in the population of field blanks (and associated environmental samples) was estimated with a confidence level (cl) of 95 percent to be below the 13th-highest ($m+1$ ranked) field blank concentration for at least 80 percent of the samples ($p = 0.80$).

Quality-Control Results

Variability

Variability was evaluated with measures of reproducibility and reliability. Reproducibility was calculated from paired measurements from replicates as normalized absolute difference (NAD) and relative standard deviation (RSD). The NADs and RSDs, as well as paired measurements with censored or estimated values for which RSDs were not calculated, were compared to specified criteria to determine if the paired measurements had acceptable reproducibility. Reliability was estimated from pooled RSDs.

Reproducibility

Reproducibility was estimated from calculations of NADs for radiochemical constituents (tables 2–4; at back of report) and RSDs for inorganic and organic constituents (tables 5–10; at back of report). (Statistical calculations were done using unrounded concentration data and, because concentrations in results tables were rounded to the least significant figure, the statistical results presented in the tables may differ slightly from statistical calculations using concentration results in the tables). Relative standard deviations were used to determine the reproducibility of inorganic and organic constituents because uncertainties, which are necessary for calculating NADs, were not provided with results for these constituents. The calculated NADs and RSDs, as well as paired measurements with censored or estimated concentrations for which RSDs were not calculated, were compared to criteria previously used by the INLPO and (or) the State of Idaho INL Oversight Program to determine if the paired measurements had acceptable reproducibility. The reproducibility for a constituent from a replicate pair was considered acceptable if:

1. the NAD was less than or equal to 1.96 (Williams, 1996, p. 14; Bartholomay and Twining, 2010, p. 14–15),

2. the RSD was less than 14 percent (this corresponds to the relative percent difference of less than 20 percent used by the Idaho National Laboratory Oversight Program [2002, p. 5–22] and Bartholomay and Twining [2010, p. 15]),

3. both measurements were censored and (or) estimated because they were less than the reporting level for that analysis (Williams, 1996, p. 15), or

4. one measurement was censored or estimated and the other measurement was within one detection limit of the larger of the estimated value or the reporting level, or the measurements were within one detection limit of each other (Idaho National Laboratory Oversight Program, 2002, p. 5–22). For results reported using the LRL as the reporting level, the detection limit was the

LT-MDL (one-half of the LRL). For results reported with the MRL as the reporting level, the detection limit was approximated as one-half of the MRL.

If the percentage of paired measurements with acceptable reproducibility for a constituent was greater than or equal to 90 percent, then the reproducibility for that constituent was considered acceptable for the period 1996–2001 (table 11, at back of report). If the percentage was less than 90 percent for a constituent, then the results for that constituent were investigated further (Idaho National Laboratory Oversight Program, 2002, p. 6–4).

There were 63 replicate pairs with measurements of gross-alpha and gross-beta radioactivity, 93 with measurements of gamma radioactivity (all gamma radionuclide results were less than reporting levels, so only cesium-137 was reported by the RESL), 204 with measurements of tritium, 123 with measurements of strontium-90, and 28 with measurements of the plutonium and americium radionuclides (tables 2–4). All these radiochemical constituents had acceptable reproducibility (that is, NAD ≤1.96) between their paired measurements except for one result for americium-241, two results for gross-beta radioactivity and cesium-137, four results for strontium-90, and eight results for tritium. NADs calculated for two replicate pairs for gross-beta radioactivity (9.10 and 8.73 in October 1996) and two replicate pairs for tritium (8.72 and 8.01 in April 2001) that did not have acceptable reproducibility were a result of switched sample bottles at the USGS laboratory at the INL or at the RESL. Measurements from switched sample bottles are easily detected if water from the two sites had large differences in concentration (incorrect results identified in this report as resulting from switched sample bottles were subsequently corrected in the USGS National Water Information System database). The percentage of paired measurements with acceptable reproducibility for each radiochemical constituent was greater than or equal to 96 percent (table 11).

There were 131, 202, 7, and 83 replicate pairs with measurements of sodium, chloride, fluoride, and sulfate, respectively, and 98 with measurements of each of the nutrient species (tables 5, 6, and 11). All paired measurements of major ions and nutrients had acceptable reproducibility except for 5 results for chloride, 7 for ammonia, 4 for nitrate + nitrite (the aquifer is an oxidizing environment, so nitrate + nitrite will hereafter be referred to as nitrate), 3 for nitrite, and 11 for orthophosphate. Two of the RSDs calculated for chloride (117 and 117 percent in April 1998; table 5) and nitrate (97 and 96 percent in October 1998; table 6) that did not have acceptable reproducibility were a result of switched sample bottles. The percentage of paired measurements with acceptable reproducibility was greater than or equal to 90 percent for each constituent except for orthophosphate (89 percent). Because relative variability generally increases as concentrations decrease, the slightly lower percentage of acceptable results for orthophosphate probably was a result of the consistently low concentrations measured for this constituent. Of 196 orthophosphate measurements from replicate pairs, 194 were less than or equal to 0.045 mg/L as phosphorous and were less than or equal to three times the reporting level for orthophosphate.

There were 97 replicate pairs with measurements of total dissolved chromium, 19 for hexavalent chromium, 5 for selenium and thallium, and 10 for all the other metals (tables 7, 8, and 11). All paired measurements of metals had acceptable reproducibility except one result each for aluminum, arsenic, manganese, and nickel, two results for zinc, four results for chromium, and nine results for hexavalent chromium. The percentage of paired measurements with acceptable reproducibility was greater than or equal to 90 percent for all of the metals except for zinc (80 percent) and hexavalent chromium (53 percent). Zinc concentrations were less than or equal to 3 µg/L for the two paired measurements for zinc without acceptable reproducibility, which was less than or equal to three times the reporting level for zinc. This indicates that, like orthophosphate, the larger relative variability for zinc was a result of zinc concentrations near the reporting level. The large variability and low reproducibility for measurements of hexavalent chromium were probably related to the inability to preserve hexavalent chromium after collection of the water sample (Rogerson and others, 1997).

There were 32 replicate pairs with measurements of VOCs (table 11), although only 1,1-dichloroethene, tetrachloroethene, tetrachloromethane, toluene, 1,1,1-trichloroethane, trichloroethene, and trichloromethane had replicate pairs with concentrations that exceeded the reporting level (table 9). For these VOCs, the reproducibility was acceptable for all the calculated RSDs except for one result each for tetrachloroethene, toluene, and 1,1,1-trichloroethane. The percentage of paired measurements with acceptable reproducibility was 97 percent for tetrachloroethene, toluene, and 1,1,1-trichloroethane and 100 percent for all other VOCs.

There were 21 replicate pairs with measurements of TOC (tables 10 and 11), and reproducibility was acceptable in 8 (38 percent) of the paired measurements. The NWQL presented laboratory QA results for organic constituents (available for December 1999–December 2001) that consistently showed variable recovery of TOC (U.S. Geological Survey, 2012b), which indicates reproducibility issues with the laboratory method.

Reliability

The reliability of radiochemical, inorganic, and organic constituents was estimated, for discrete concentration ranges, as a pooled RSD (table 12, at back of report). Reliability was estimated for discrete concentration ranges because (1) pooled RSDs should be calculated from samples with similar variability (Taylor, 1987, p. 22) and (2) variability and RSDs are a function of concentration (Martin, 2002, p. 35). Generally, variability and RSDs decrease as concentrations increase. Qualitatively, and using the criterion specified for reproducibility, pooled RSDs less than 14 percent indicated

that the measurements for that constituent and concentration range met a minimum objective for reliability. However, pooled RSDs provide a precise measure of reliability (which increases as pooled RSDs decrease) that can be used to calculate confidence limits for water-quality measurements (Martin, 2002, p. 50–51).

The RSDs calculated for replicate pairs were used to identify appropriate concentration ranges for each constituent to evaluate reliability with pooled RSDs (relative standard deviations were calculated for radiochemical constituents to calculate the pooled RSDs and were calculated for a replicate pair only if both radiochemical concentrations equaled or exceeded the method detection limit and the minimum detectable concentration (MDC) of three times the CSU [Mann, 1996, p. 33–36]). Discrete concentration ranges were selected for each constituent by plotting the RSD and mean constituent concentration for each replicate pair and grouping concentration ranges based on differences in the ranges of plotted RSDs. For example, figure 5 shows a plot of RSDs and mean sodium concentrations from replicate pairs. The range of RSDs was largest, 0 to 13 percent, for a sodium concentration range of 5 mg/L (the lowest concentration measured) to less than 30 mg/L. A smaller range of RSDs, 0–4.0 percent, was calculated for a concentration range of 30–180 mg/L, and an RSD of 1.3 percent was calculated from a replicate pair with a mean sodium concentration of 668 mg/L (table 12).

Relative standard deviations were not calculated for gross-alpha radioactivity, cesium-137, plutonium-238, plutonium-239+240, and americium-241 because none of the replicate pairs had concentrations that exceeded the MDC for these constituents. Relative standard deviations were calculated from 7 replicate pairs for gross-beta radioactivity, 74 for tritium, and 25 for strontium-90. Pooled RSDs for gross-beta radioactivity were 18 percent for the concentration range 6.0 to 12 pCi/L as Cs-137 and 17 percent for the concentration range 50 to 60 pCi/L as Cs-137 (table 12). The low concentration range was within three times the method detection limit of 4 pCi/L for gross-beta radioactivity (Mann, 1996, p. 35); therefore, the high variability and low measurement reliability at these concentrations was a reasonable result. Pooled RSDs for tritium were calculated for three concentration ranges: 500 to less than 2,000, 2,000 to less than 20,000, and 20,000 to 80,000 pCi/L. The pooled RSDs for tritium, 8.4, 7.2, and 1.3 percent, respectively, decreased as concentrations increased, and indicated a low variability and high reliability for measurements of tritium across all concentration ranges. Pooled RSDs for strontium-90 of 6.1 and 3.4 percent were calculated for concentration ranges of 12 to less than 25 and 25 to 250 pCi/L, which indicated a low variability and high reliability for measurements of strontium-90 at these concentrations. A pooled RSD for strontium-90 of 23 percent was calculated for the concentration range 6.0 to less than 12 pCi/L. The low measurement reliability for this concentration range was attributed to concentrations near the method detection limit of 5 pCi/L (Mann, 1996, p. 35).

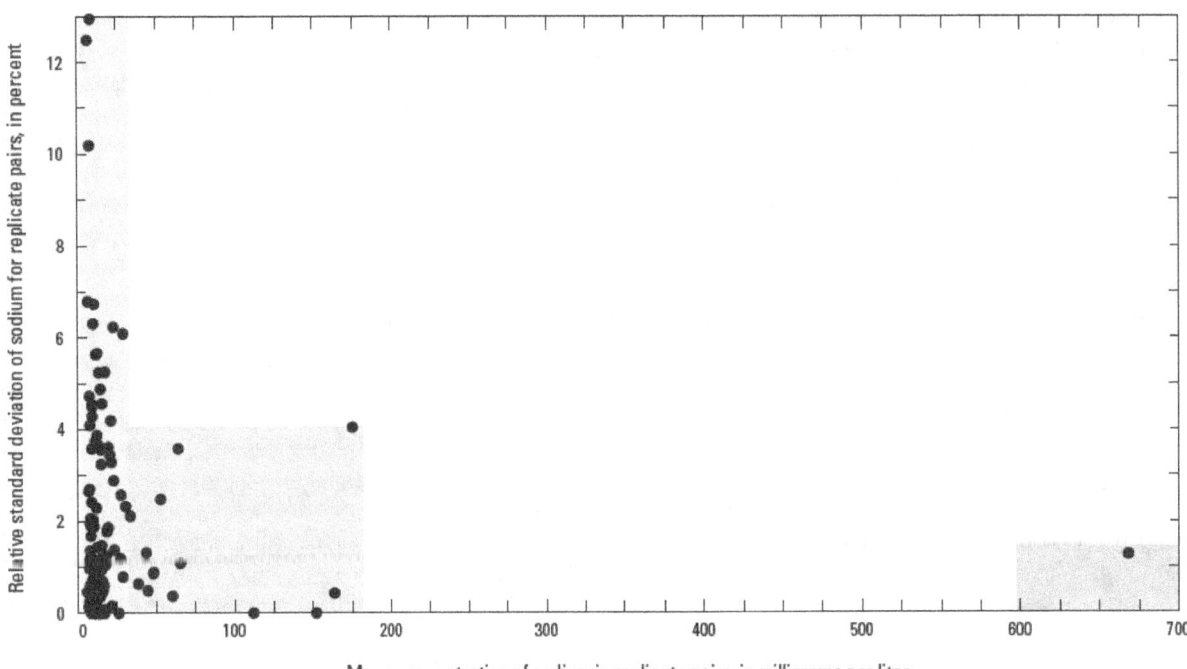

Figure 5.   Variability of sodium as a function of sodium concentration. Gray rectangles indicate three discrete concentration ranges with different ranges of variability (shown as relative standard deviations).

Pooled RSDs for the major ions (from RSDs calculated from 131 replicate pairs for sodium, 202 for chloride, 7 for fluoride, and 83 for sulfate) ranged from 1.3 to 3.2 percent for sodium, 1.5 to 14 percent for chloride, 6.6 percent for fluoride, and 1.5 to 1.8 percent for sulfate. The pooled RSD for chloride concentrations ranging from 20 to less than 60 mg/L was 14 percent. This result was calculated from 43 replicate pairs, and 41 of the calculated RSDs were less than or equal to 3.6 percent. Additionally, the pooled RSD for chloride concentrations ranging from 60 to 350 mg/L, 1.5 percent, did not include results from two replicate pairs with switched sample bottles (these results were excluded from the pooled RSD calculations because any environmental samples with a similar large difference in concentration because of switched sample bottles, when compared to the long record of historical concentrations for the particular site, would be easily detected). The pooled RSDs for the major ions indicate that measurements of the major ions had low variability and high reliability.

Pooled RSDs for nutrients were determined from RSDs calculated from 16 replicate pairs for ammonia, 98 for nitrate, 7 for nitrite, and 55 for orthophosphate (table 12). One RSD each was calculated for ammonia and orthophosphate from a mean replicate concentration that exceeded three times their reporting levels. The RSDs, 0.9 and 7.6 percent for ammonia and orthophosphate, respectively, showed low variability and high reliability for these measurements. For concentrations near their reporting levels, pooled RSDs for ammonia, nitrite, and orthophosphate were 23, 17, and 16

percent, respectively, and showed low measurement reliability. Pooled RSDs for nitrate + nitrite, for concentration ranges of 0.3 to less than 0.7, 0.7 to 6.0, and 10 to 60 mg/L as nitrogen (N), were 2.0, 3.9, and 17 percent, respectively. The pooled RSD of 3.9 percent was calculated after excluding results from two replicate pairs with switched sample bottles, and the pooled RSD of 17 percent was calculated from two replicate pairs with RSDs less than or equal to 1.3 percent and one replicate pair with an RSD of 88 percent. Except for a few measurements, nitrate showed low variability and high reliability.

The pooled RSDs for metals were calculated from results of 1 to 4 replicate pairs for antimony, cadmium, cobalt, copper, lead, manganese, nickel, and selenium; 8 to 11 replicate pairs for aluminum, arsenic, barium, molybdenum, uranium, zinc, and hexavalent chromium; and 38 replicate pairs for chromium. Antimony, barium, cadmium, cobalt, copper, lead, manganese, molybdenum, and uranium had pooled RSDs of less than or equal to 6.4, indicating low variability and high reliability for measurements of these metals. Pooled RSDs for aluminum and arsenic were 12 percent, and for nickel and selenium were 22 and 15 percent, respectively. The high variability and low reliability for measurements of nickel and selenium were attributed to measured concentrations near their reporting levels. Pooled RSDs were calculated for two concentration ranges of zinc and hexavalent chromium and three concentration ranges of chromium. The low concentration range for zinc, 1.0 to 3.0 µg/L, was near the reporting level of 1 µg/L for zinc and had a pooled RSD of 33 percent. At the high concentration range for zinc, 10 to 400 µg/L, the pooled RSD of 1.2 percent indicates high measurement reliability. The three concentration ranges for chromium, 2.0 to less than 13, 13 to 40, and 70 to 200 µg/L had pooled RSDs of 11, 5.1, and 1.2 percent, respectively. This indicates that reliability of chromium measurements was high across all concentration ranges. The reliability of measurements of hexavalent chromium was low, with pooled RSDs for the low and high concentration ranges of 40 and 51 percent. The high variability for hexavalent chromium probably was related to the inability to preserve hexavalent chromium after collection of the water sample (see section "Quality Control Results").

Pooled RSDs for the seven VOCs with replicate pair concentrations greater than reporting levels were calculated from 16 or fewer replicate pairs. The pooled RSDs for 1,1-dichloroethene, tetrachloromethane, 1,1,1-trichloroethane, trichloroethene, and trichloromethane were less than or equal to 5.6 percent, indicating low variability and high reliability for measurements of these VOCs. The pooled RSD was 21 percent for tetrachloroethene and 30 percent for toluene. The higher variability for these VOCs was attributed to their low measured concentrations that were at or near their respective reporting levels.

The pooled RSD for total organic carbon (TOC), calculated from results of 13 replicate pairs where both TOC concentrations exceeded the reporting level, was 60 percent. The NWQL presented laboratory QA results for organic constituents (available for December 1999–December 2001) that consistently showed variable recovery of TOC (U.S. Geological Survey, 2012b). Consequently, the high variability and low reliability of TOC measurements was attributed to variability in the laboratory method. Because the laboratory method was unreliable during the study period, the INLPO will evaluate future QC results for TOC to determine if collection of TOC samples should be continued.

Bias

Bias was estimated from field, equipment, and source solution blanks. Field blanks were collected to estimate the potential contamination bias of selected constituents in environmental samples caused by the preservation, storage, handling, shipping, processing, and analysis of the blanks and samples. The field blanks did not include any contamination bias from equipment, such as the dedicated pumps and casing present in most wells or the portable sampling equipment used at 21 wells (Mann, 1996, p. 21–29; Bartholomay and others, 2003, p. 26–34). Equipment blanks were used to estimate the potential bias of selected constituents from cross contamination of samples collected with portable sampling equipment. A source-solution blank was collected to confirm that the deionized water (DIW) used as a source solution for some blanks was free of the constituents of interest.

The criterion used to determine when a detectable concentration from a blank was due to inadvertent sample bias, rather than instrument background uncertainty (sometimes referred to as "noise"), was a concentration exceeding the method detection limit of 3 times the CSU for radiochemical constituents (Mann, 1996, p. 33–36) or a concentration exceeding the reporting level for inorganic and organic constituents. When a detectable constituent concentration was measured in a field or equipment blank sample, and if more than one field or one equipment blank result was available for that constituent, order statistics and the binomial probability distribution were used to estimate the potential contamination bias of the constituent in environmental samples. These statistical methods were used to calculate with 91, 93, or 95 percent confidence that contamination bias in 70 or 80 percent of the water-quality measurements for a constituent was less than a specific concentration. The potential contamination bias of the constituent also was estimated as the percent of the lowest environmental concentration potentially affected. Using the lowest environmental concentration potentially affected provided a worst-case estimate of potential contamination bias.

Source-Solution Blank

The source-solution blank was collected on October 28, 1996, to verify that the DIW used as a source solution for equipment and field blanks was free of the constituents of interest. The source-solution blank was analyzed for tritium, strontium-90, cesium-137, sodium, chloride, sulfate, chromium, and hexavalent chromium (table 13). The only constituent with a detectable concentration was hexavalent chromium, which had a measured concentration of 2 µg/L. The hexavalent chromium in the source-solution blank may have been present in the source solution, acquired during the preservation, storage, handling, shipping, processing, and analysis of the sample, or may be an artifact of high variability in the analytical method. Of the 117 measurements of hexavalent chromium from environmental samples in 1996, 98 measurements were less than 20 µg/L. Consequently, if the positive bias was from a source other than the source solution, a bias of 2 µg/L could produce a bias of 10 percent or more in about 84 percent of the hexavalent chromium measurements.

Field Blanks

Fourteen field blanks were collected from 1996 to 2001 (table 13). The field blanks collected at sample sites included sample bottles for the same constituents as were collected for the environmental samples. Analysis of field blanks included tritium and strontium-90 from all 14 blanks, chloride from 13 blanks, sodium and sulfate from 6 blanks, cesium-137 and chromium from 5 blanks, and hexavalent chromium and ammonia from 1 blank (the field blank collected on April 20, 1999, also included most of the remaining constituents discussed in this report, all of which were nondetects). A detectable concentration was measured for only two constituents. Chloride was detected in the field blank from April 25, 1996, at a concentration of 0.2 mg/L and ammonia was detected in the field blank from April 20, 1999, at a concentration of 0.03 mg/L as N. The source of chloride in the field blank could have been the DIW source solution or some unidentified source during the storage, handling, shipping, processing, and analysis of the sample. The source of ammonia may have occurred at the NWQL, because the laboratory shows a positive bias for ammonia in late-April 1999 (U.S. Geological Survey, 2012c), the period of time when the sample was analyzed.

The potential bias of chloride in all environmental samples was estimated using order statistics for the chloride concentrations from field blanks and the binomial probability distribution. The order ranking for the chloride measurements used the LT-MDL (<0.145 and <0.15) instead of the LRL (<0.29 and <0.30) for measurements from field blanks collected in 2000 and 2001 because these measurements, although reported as less than LRL, were actually less than LT-MDL (Childress and others, 1999, p. 8–9). There were 13 chloride measurements from field blanks, and the 13th (or $m+1$) ranked concentration (that is, the maximum measured concentration from the blanks) was 0.2 mg/L. Using a p-value of 0.8 (probability of success in table 14), it was estimated with a confidence level (cl) of 95 percent that at least 80 percent ($100p$ percent) of the environmental samples had a chloride bias of less than 0.2 mg/L (table 14). Because the minimum concentration from the 1,987 environmental chloride concentrations measured was 2.6 mg/L, there is 95 percent confidence that at least 80 percent of the environmental samples had a potential bias of less than 8 percent.

In April 1999, 43 environmental samples were collected that included measurements of ammonia and that could have been affected by a potential ammonia bias at the NWQL. Of these 43 samples, 35 measurements were less than the reporting level and did not have any detectable positive bias for ammonia. The other eight ammonia measurements ranged from 0.02 to 0.55 mg/L of N, and a potential bias of 0.03 mg/L as N would produce a bias of ammonia of as much as 5 to 150 percent in these measurements.

Equipment Blanks

Twelve equipment blanks were collected (table 13). The equipment blanks included sample bottles for the same constituents as were collected for environmental samples at the site where the blank was collected. Tritium, strontium-90, and chloride were analyzed from all 12 equipment blanks; sodium, sulfate, and chromium were analyzed from 7

equipment blanks; cesium-137 was analyzed from 4 equipment blanks; and hexavalent chromium was analyzed from 3 equipment blanks. The radionuclides and chromium were not detected in any of the blanks, chloride was detected in four blanks, sodium in three blanks, and sulfate and hexavalent chromium were each detected in one blank. These detectable concentrations were not correlated with a specific type of equipment or sampling personnel, but three of the four chloride detections were from equipment blanks collected following collection of environmental samples at well PW 1 (PW 1 had measured chloride concentrations of 133 to 386 mg/L during 1996–2001).

Hexavalent chromium was detected in an equipment blank collected on October 28, 1996, at a concentration of 2 µg/L. The same concentration was measured for the source-solution blank collected on the same day as the equipment blank, which indicates that the hexavalent chromium detected in the equipment blank probably is not from the portable sampling equipment.

The potential bias of chloride in environmental samples collected with portable sampling equipment was estimated from 12 chloride measurements from equipment blanks. The 12th (or $m+1$) ranked concentration (that is, the maximum measured concentration from the blanks) was 0.55 mg/L. Using a p-value of 0.8 (probability of success in table 14), it was estimated with a confidence level (cl) of 93 percent that at least 80 percent ($100p$ percent) of the environmental samples had a chloride bias of less than 0.55 mg/L (table 14). Because the minimum chloride concentration from 365 chloride concentrations measured from environmental samples collected using portable sampling equipment was 3.1 mg/L, there is 93 percent confidence that at least 80 percent of these environmental samples had a potential bias of less than 18 percent.

The potential bias of sodium and sulfate in environmental samples collected with portable sampling equipment was estimated from seven measurements of each constituent from equipment blanks. The maximum measured concentration from the equipment blanks for sodium and sulfate was 0.73 and 0.57 mg/L, respectively. Using a p-value of 0.7, it was estimated with a confidence level of 91 percent that at least 70 percent of these environmental samples had a sodium bias of less than 0.73 mg/L and a sulfate bias of less than 0.57 mg/L. From 149 and 209 measurements of sodium and sulfate, respectively, the minimum sodium and sulfate concentrations measured from environmental samples collected using portable sampling equipment were 6.1 for sodium and 13 mg/L for sulfate (table 14). Consequently, there is 91 percent confidence that at least 70 percent of these environmental samples had a potential bias of less than 12 percent for sodium and less than 5 percent for sulfate.

Summary and Conclusions

The U.S. Geological Survey, in cooperation with the U.S. Department of Energy, has been studying the water quality of the eastern Snake River Plain aquifer at and near the Idaho National Laboratory (INL) since 1949. The INL extends over about 890 square miles of the eastern Snake River Plain in southeastern Idaho and overlies about 8 percent of the fractured basalt sole-source aquifer. The U.S. Geological Survey began routine collection of water-quality samples in 1964 to monitor the concentrations and delineate the movement of radiochemical and chemical wastes discharged to the subsurface at the INL.

Beginning in 1980, quality-control samples were routinely collected at groundwater and surface water sites to ensure and document the quality of the environmental data. Quality-control samples collected from 1996 to 2001 included 204 replicates and 27 blanks. Measurements from replicates were used to calculate the variability (as reproducibility and reliability) of environmental measurements of radiochemical, inorganic, and organic constituents due to sample collection and analysis. Measurements from field and equipment blanks were used to estimate the potential bias of selected constituents from (1) sample collection and analysis of environmental samples and (2) from cross contamination of environmental samples collected with portable sampling equipment.

Variability was calculated from paired measurements from replicates as the normalized absolute difference (NAD) for radiochemical constituents and the relative standard deviation (RSD) for inorganic and organic constituents. The NADs and RSDs, as well as paired measurements with censored or estimated concentrations for which RSDs were not calculated, were compared to specified criteria to determine if the paired measurements had acceptable reproducibility. If the percentage of paired measurements with acceptable reproducibility for a constituent was greater than or equal to 90 percent, then the reproducibility for that constituent was considered acceptable for the period 1996–2001. The percentage of paired measurements with acceptable reproducibility was greater than 90 percent for all of the constituents except orthophosphate (89 percent), zinc (80 percent), hexavalent chromium (53 percent), and total organic carbon (TOC) (38 percent). The low reproducibility for orthophosphate and zinc were attributed to calculation of RSDs from replicates with low concentrations of these constituents. The low reproducibility for hexavalent chromium was attributed to the inability to preserve hexavalent chromium after collection of the water sample, and the low reproducibility for TOC was attributed to high variability with the analytical method.

The reliability of radiochemical, inorganic, and organic measurements was estimated from pooled RSDs that were calculated for discrete concentration ranges for each constituent. For most constituents, pooled RSDs were inversely correlated with concentration and, as a result, pooled RSDs often were higher and reliability was lower at low concentrations. For example, pooled RSDs of 15–33 percent were calculated for low concentrations of gross-beta radioactivity, strontium-90, ammonia, nitrite, orthophosphate, nickel, selenium, zinc, tetrachloroethene, and toluene. Low pooled RSDs of 0–12 percent were calculated for all other concentration ranges of these constituents, and for all other constituents, except for one concentration range for gross-beta radioactivity, chloride, and nitrate + nitrite; two concentration ranges for hexavalent chromium; and for TOC. Pooled RSDs for the 50–60 picocuries per liter concentration range of gross-beta radioactivity (reported as cesium-137) and the 10–60 milligrams per liter (mg/L) concentration range of nitrate + nitrite (reported as nitrogen [N]) were 17 percent. Chloride had a pooled RSD of 14 percent for the 20 to less than 60 mg/L concentration range. High pooled RSDs of 40 and 51 percent were calculated for two concentration ranges for hexavalent chromium and of 60 percent for TOC.

Measurements from (1) field blanks were used to estimate the potential bias associated with environmental samples from sample collection and analysis, (2) equipment blanks were used to estimate the potential bias from cross contamination of samples collected from wells where portable sampling equipment was used, and (3) a source-solution blank were used to verify that the deionized water source-solution was free of the constituents of interest. If more than one measurement was available, the bias was estimated using order statistics and the binomial probability distribution. The source-solution blank had a detectable concentration of hexavalent chromium of 2 micrograms per liter. If this bias was from a source other than the source solution, then about 84 percent of the 117 hexavalent chromium measurements from environmental samples could have a bias of 10 percent or more. Fourteen field blanks were collected, and only chloride (0.2 mg/L) and ammonia (0.03 mg/L as N), in one blank each, had detectable concentrations. It was estimated with a confidence level of 95 percent that at least 80 percent of the 1,987 chloride concentrations measured from all environmental samples had a potential chloride bias of less than 8 percent. The ammonia bias, which may have occurred at the analytical laboratory, could produce a potential bias of 5–150 percent in eight potentially affected ammonia measurements. Twelve equipment blanks were collected, chloride was detected in four of these blanks, sodium in three blanks, and sulfate and hexavalent chromium were each detected in one blank. The concentration of hexavalent chromium in the equipment blank was the same concentration as in the source-solution blank collected on the same day, which indicates that the hexavalent chromium in the equipment blank is probably from a source other than the portable sampling equipment; for example, from the sample bottles or the source-solution water itself. The potential bias for chloride, sodium, and sulfate measurements was estimated for environmental samples collected using portable sampling equipment. For chloride, it was estimated with a confidence level of 93 percent that at least 80 percent of the measurements had a chloride bias of less than 18 percent. For sodium and sulfate, it was estimated with a confidence level of 91 percent that at least 70 percent of the measurements had a sodium or sulfate bias of less than 12 and 5 percent, respectively.

References Cited

Barnett, P.R., and Mallory, E.C., 1971, Determination of minor elements in water by emission spectroscopy: U.S. Geological Survey Techniques of Water-Resources Investigations, book 5, chap. A2, 31 p. (Also available at http://pubs.er.usgs.gov/publication/twri05A2.)

Bartholomay, R.C., Knobel, L.L., and Rousseau, J.P., 2003, Field methods and quality-assurance plan for quality-of-water activities, U.S. Geological Survey, Idaho National Engineering and Environmental Laboratory, Idaho: U.S. Geological Survey Open-File Report 2003–42 (DOE/ID-22182), 45 p. (Also available at http://pubs.er.usgs.gov/publication/ofr0342.)

Bartholomay, R.C., and Twining, B.V., 2010, Chemical constituents in groundwater from multiple zones in the eastern Snake River Plain aquifer at the Idaho National Laboratory, 2005–2008: U.S. Geological Survey Scientific Investigations Report 2010–5116 (DOE/ID-22211), 82 p. (Also available at http://pubs.er.usgs.gov/publication/sir20105116.)

Bodnar, L.Z., and Percival, D.R., eds., 1982, Analytical Chemistry branch procedures manual—Radiological and Environmental Services Laboratory: U.S. Department of Energy Report IDO-12096, variously paged.

Childress, C.J.O., Foreman, W.T., Conner, B.F., and Maloney, T.J., 1999, New reporting procedures based on long-term method detection levels and some considerations for interpretations of water-quality data provided by the U.S. Geological Survey National Water Quality Laboratory: U.S. Geological Survey Open-File Report 99–193, 19 p. (Also available at http://pubs.er.usgs.gov/publication/ofr99193.)

Faires, L.M., 1992, Methods of analysis by the U.S. Geological Survey National Water Quality Laboratory—Determinations of metals in water by inductively coupled plasma-mass spectroscopy: U.S. Geological Survey Open-File Report 92–634, 28 p. (Also available at http://pubs.er.usgs.gov/publication/ofr92634.)

Fishman, M.J., ed., 1993, Methods of analysis by the U.S. Geological Survey National Water Quality Laboratory—Determination of inorganic and organic constituents in water and fluvial sediments: U.S. Geological Survey Open-File Report 93–125, 217 p. (Also available at http://pubs.er.usgs.gov/publication/ofr93125.)

Fishman, M.J., and Friedman, L.C., eds., 1989, Methods of determination of inorganic substances in water and fluvial sediments (3d ed.): U.S. Geological Survey Techniques of Water-Resources Investigations, book 5, chap. A1, 545 p. (Also available at http://pubs.er.usgs.gov/publication/twri05A1.)

Friedman, L.C., and Erdmann, D.E., 1982, Quality assurance practices for the chemical and biological analyses of water and fluvial sediments: U.S. Geological Survey Techniques of Water-Resources Investigations, book 5, chap. A6, 181 p, (Also available at http://pubs.er.usgs.gov/publication/twri05A6.)

Goerlitz, D.F., and Brown, Eugene, 1972, Methods for analysis of organic substances in water: U.S. Geological Survey Techniques of Water-Resources Investigations, book 5, chap. A3, 40 p. (Also available at http://pubs.er.usgs.gov/publication/twri05A3_1972.)

Idaho National Laboratory Oversight Program, 2002, 2001 environmental surveillance report—A compilation and explanation of data collected by the INEEL Oversight Program during 2001: State of Idaho, variously paged.

Kateman, G., and Buydens, L., 1993, Quality control in analytical chemistry: New York, John Wiley and Sons, Inc., 317 p.

Knobel, L.L., Bartholomay, R.C., and Rousseau, J.P., 2005, Historical development of the U.S. Geological Survey Hydrologic Monitoring and Investigative Programs at the Idaho National Engineering and Environmental Laboratory, Idaho, 1949 to 2001: U.S. Geological Survey Open-File Report 2005–1223 (DOE/ID-22195), 93 p. (Also available at http://pubs.er.usgs.gov/publication/ofr20051223.)

Knobel, L.L., Bartholomay, R.C., Tucker, B.J., and Williams, L.M., 1999a, Chemical and radiochemical constituents in water from wells in the vicinity of the Naval Reactors Facility, Idaho National Engineering and Environmental Laboratory, Idaho, 1996: U.S. Geological Survey Open-File Report 99–272 (DOE/ID-22160), 58 p. (Also available at http://pubs.er.usgs.gov/publication/ofr99272.)

Knobel, L.L., Bartholomay, R.C., Tucker, B.J., Williams, L.M., and Cecil, L.D., 1999b, Chemical constituents in ground water from 39 selected sites with an evaluation of quality assurance data, Idaho National Engineering and Environmental Laboratory and vicinity, Idaho: U.S. Geological Survey Open-File Report 99–246 (DOE/ID-22159), 58 p. (Also available at http://pubs.er.usgs.gov/publication/ofr99246.)

Ludtke, A.S., Woodworth, M.T., and Marsh, P.S., 1999, Quality-assurance results for routine water analyses in U.S. Geological Survey laboratories, water year 1997: U.S. Geological Survey Water-Resources Investigations Report 99–4057, 186 p. (Also available at http://pubs.er.usgs.gov/ publication/wri994057.)

Ludtke, A.S., Woodworth, M.T., and Marsh, P.S., 2000, Quality-assurance results for routine water analyses in U.S. Geological Survey laboratories, water year 1998: U.S. Geological Survey Water-Resources Investigations Report 2000–4176, 198 p. (Also available at http://pubs.er.usgs. gov/publication/wri20004176.)

Mann, L.J., 1996, Quality-assurance plan and field methods for quality-of-water activities, U.S. Geological Survey, Idaho National Engineering Laboratory, Idaho: U.S. Geological Survey Open-File Report 96–615 (DOE/ID-22132), 37 p. (Also available at http://pubs. er.usgs.gov/publication/ofr96615.)

Martin, J.D., 2002, Variability of pesticide detections and concentrations in field replicate water samples collected for the National Water-Quality Assessment Program, 1992–97: U.S. Geological Survey Water-Resources Investigations Report 2001–4178, 84 p. (Also available at http://pubs. er.usgs.gov/publication/wri20014178.)

McCurdy, D.E, Garbarino, J.R., and Mullin, A.H., 2008, Interpreting and reporting radiological water-quality data: U.S. Geological Survey Techniques and Methods, book 5, chap. B6, 33 p. (Also available at http://pubs.er.usgs.gov/ publication/tm5B6.)

Mueller, D.K., 1998, Quality of nutrient data from streams and ground water sampled during 1993–95—National Water-Quality Assessment Program: U.S. Geological Survey Open-File Report 98–276, 25 p. (Also available at http:// pubs.er.usgs.gov/publication/ofr98276.)

Nace, R.L., Stewart, J.W., Walton, W.C, and others, 1959, Geography, geology, and water resources of the National Reactor Testing Station, Idaho. Part 3—Hydrology and water resources: U.S. Atomic Energy Commission, Idaho Operations Publication IDO-22033-USGS, 253 p.

Olmstead, F.H., 1962, Chemical and physical character of ground water in the National Reactor Testing Station, Idaho: U.S. Atomic Energy Commission, Idaho Operations Office Publication IDO-22043-USGS, 142 p.

Parr, J.G., and Porterfield, D.R., 1997, Evaluation of radiochemical data usability: U.S. Department of Energy, Office of Environmental Management, es/er/ms-5, 67 p.

Pritt, J.W., and Raese, J.W., 1995, Quality assurance/quality control manual—National Water Quality Laboratory: U.S. Geological Survey Open-File Report 95-443, 35 p. (Also available at http://pubs.er.usgs.gov/publication/ofr95443.)

Rattray, G.W., and Campbell, L.J., 2003, Radiochemical and chemical constituents in water from selected wells and springs from the southern boundary of the Idaho National Engineering and Environmental Laboratory to the Hagerman area, Idaho, 2002: U.S. Geological Survey Open-File Report 2004–1004 (DOE/ID-22190), 22 p. (Also available at http://pubs.er.usgs.gov/publication/ ofr20041004.)

Robertson, J.B., Schoen, R., and Barraclough, J.T., 1974, The influence of liquid waste disposal on the geochemistry of the National Reactor Testing Station, Idaho, 1952–1970: U.S. Geological Survey Open-File Report IDO-22053 (73–238), 231 p. (Also available at http://pubs.er.usgs.gov/ publication/ofr73238.)

Rogerson, P.F., Ardourel, Harold, and White, Ralph, 1997, Replacement or elimination of NWQL procedures: U.S. Geological Survey National Water-Quality Laboratory Technical Memorandum 1997.09, unpaged, accessed November 6, 2012, at http://wwwnwql.cr.usgs.gov/dyn. shtml?techmemo.

Rose, D.L., and Schroeder, M.P., 1995, Methods of analysis by the U.S. Geological Survey National Water Quality Laboratory—Determination of volatile organic compounds in water by purge and trap capillary gas chromatograph-mass spectrometry: U.S. Geological Survey Open-File Report 94–708–W, 26 p. (Also available at http://pubs. er.usgs.gov/publication/ofr94708W.)

Skougstad, M.W., Fishman, M.J., Friedman, L.C., Erdmann, D.E., and Duncan, S.S., 1979, Methods for determination of inorganic substances in water and fluvial sediments: U.S. Geological Survey Techniques of Water-Resources Investigations (2d ed.), book 5, chap. A1, 626 p. (Also available at http://pubs.er.usgs.gov/publication/ twri05A1_1979.)

Spiegel, M.R., 1998, Schaum's outline of theory and problems of probability and statistics: New York, McGraw-Hill, 372 p.

Taylor, J.K., 1987, Quality assurance of chemical measurements: Chelsea, Mich., Lewis Publishers, Inc., 328 p.

Thatcher, L.L., Janzer, V.J., and Edwards, K.W., 1977, Methods for the determination of radioactive substances in water and fluvial sediments: U.S. Geological Survey Techniques of Water-Resources Investigations, book 5, chap. A5, 95 p. (Also available at http://pubs.er.usgs.gov/publication/twri05A5.)

Timme, P.J., 1995, National Water Quality Laboratory, 1995 services catalog: U.S. Geological Survey Open-File Report 95–352, 120 p. (Also available at http://pubs.er.usgs.gov/publication/ofr95352.)

U.S. Department of Energy, 1995, Radiochemistry manual, revision 10: Idaho Falls, Idaho, U.S. Department of Energy, Radiological and Environmental Services Laboratory, variously paged.

U.S. Geological Survey, 2006, Collection of water samples (ver. 2.0, revised September 2006): U.S. Geological Survey Techniques of Water-Resources Investigations, book 9, chap. A4. (Also available at http://pubs.er.usgs.gov/publication/twri09A4.)

U.S. Geological Survey, 2012a, USGS Inorganic Blind Sample Project (IBSP)—Monitoring and evaluating the National Water Quality Laboratory's Inorganic Analytical Data Quality: U.S. Geological Survey database, accessed September 27, 2012, at http://bqs.usgs.gov/ibsp/index.shtml.

U.S. Geological Survey, 2012b, USGS Organic Blind Sample Project: U.S. Geological Survey database, accessed September 28, 2012, at http://bqs.usgs.gov/obsp/rp.html.

U.S. Geological Survey, 2012c, USGS Inorganic Blind Sample Project (IBSP)—QADATA summaries: U.S. Geological Survey database accessed September 28, 2012, at http://bqs.usgs.gov/ibsp/qadata.shtml.

Wegner, S.J., 1989, Selected quality assurance data for water samples collected by the U.S. Geological Survey, Idaho National Engineering Laboratory, Idaho, 1980 to 1988: U.S. Geological Survey Water-Resources Investigations Report 89–4168 (DOE/ID-22085), 91 p. (Also available at http://pubs.er.usgs.gov/publication/wri894168.)

Wershaw, R.L., Fishman, M.J., Grabbe, R.R., and Lowe, L.E., eds., 1987, Methods for the determination of organic substances in water and fluvial sediments: U.S. Geological Survey Techniques of Water-Resources Investigations, book 5, chap. A3, 80 p. (Also available at http://pubs.er.usgs.gov/publication/twri05A3.)

Williams, L.M., 1996, Evaluation of quality assurance/quality control data collected by the U.S. Geological Survey for water-quality activities at the Idaho National Engineering Laboratory, Idaho, 1989 through 1993: U.S. Geological Survey Water-Resources Investigations Report 96–4148 (DOE/ID-22129), 116 p. (Also available at http://pubs.er.usgs.gov/publication/wri964148.)

Williams, L.M., 1997, Evaluation of quality assurance/quality control data collected by the U.S. Geological Survey for water-quality activities at the Idaho National Engineering Laboratory, Idaho, 1994 through 1995: U.S. Geological Survey Water-Resources Investigations Report 97–4058 (DOE/ID-22136), 87 p. (Also available at http://pubs.er.usgs.gov/publication/wri974058.)

Williams, L.M., Bartholomay, R.C., and Campbell, L.J., 1998, Evaluation of quality-assurance/quality-control data collected by the U.S. Geological Survey from wells and springs between the southern boundary of the Idaho National Engineering Laboratory and the Hagerman area, Idaho, 1989 through 1995: U.S. Geological Survey Water-Resources Investigations Report 98–4206 (DOE/ID-22150), 83 p. (Also available at http://pubs.er.usgs.gov/publication/wri984206.)

Table 1. Reporting levels and reporting level codes for constituents analyzed by the U.S. Geological Survey National Water Quality Laboratory, 1996–2001.

[**Abbreviations:** mg/L, milligrams per liter, N, nitrogen, P, phosphorous, µg/L, micrograms per liter, MRL, minimum reporting level, LRL, laboratory reporting level]

Constituent	Dates	Reporting level	Reporting level code
Sodium (mg/L)	01-01-96–12-22-97	0.2	MRL
	12-12-97–09-30-98	0.1	MRL
	10-01-98–09-30-99	0.06	LRL
	10-01-99–10-17-00	0.09	LRL
	10-18-00–09-30-01	0.06	LRL
	10-01-01–12-31-01	0.09	LRL
Chloride (mg/L)	01-01-96–09-30-99	0.1	MRL
	10-01-99–11-02-00	0.29	LRL
	11-03-00–09-30-01	0.08	LRL
	10-01-01–12-31-01	0.33	LRL
Fluoride (mg/L)	01-01-96–10-15-00	0.1	MRL
	10-16-00–09-30-01	0.16	LRL
	10-01-01–12-31-01	0.11	LRL
Sulfate (mg/L)	01-01-96–09-30-99	0.1	MRL
	10-01-99–11-02-00	0.31	LRL
	11-03-00–12-31-01	0.11	LRL
Ammonia (mg/L as N)	01-01-96–11-13-97	0.015	MRL
	11-14-97–10-03-00	0.02	MRL
	10-04-00–12-31-01	0.041	LRL
Nitrate + nitrite (mg/L as N)	01-01-96–10-03-00	0.05	MRL
	10-04-00–12-31-01	0.047	LRL
Nitrite (mg/L as N)	01-01-96–09-30-00	0.01	MRL
	10-01-00–09-30-01	0.006	LRL
	10-01-01–12-31-01	0.008	LRL
Orthophosphate (mg/L as P)	01-01-96–10-03-00	0.01	MRL
	10-04-00–12-31-01	0.018	LRL
Aluminum (µg/L)	01-01-96–12-31-01	1	MRL
Antimony (µg/L)	01-01-96–09-30-00	1	MRL
	10-01-00–12-31-01	0.048	LRL
Arsenic (µg/L)	01-01-96–09-30-99	1	MRL
	10-01-99–09-30-01	2	LRL
	10-01-01–12-31-01	1.8	LRL
Barium (µg/L)	01-01-96–12-31-01	1	MRL
Beryllium (µg/L)	01-01-96–09-26-00	1	MRL
	10-01-00–12-31-01	0.06	LRL
Cadmium (µg/L)	01-01-96–09-30-00	1	MRL
	10-01-00–12-31-01	0.037	LRL
Cobalt (µg/L)	01-01-96–09-30-00	1	MRL
	10-01-00–12-31-01	0.015	LRL
Copper (µg/L)	01-01-96–09-30-00	1	MRL
	10-01-00–12-31-01	0.12	LRL
Lead (µg/L)	01-01-96–09-30-00	1	MRL
	10-01-00–12-31-01	0.08	LRL

Table 1.   Reporting levels and reporting level codes for constituents analyzed by the U.S. Geological Survey National Water Quality Laboratory, 1996–2001.—Continued

[**Abbreviations:** mg/L, milligrams per liter, N, nitrogen, P, phosphorous, µg/L, micrograms per liter, MRL, minimum reporting level, LRL, laboratory reporting level]

Constituent	Dates	Reporting level	Reporting level code
Manganese (µg/L)	01-01-96–09-30-00	1	MRL
	10-01-00–12-31-01	0.1	MRL
Mercury (µg/L)	01-01-96–09-30-99	0.1	MRL
	10-01-99–03-31-01	0.23	LRL
	04-01-01–09-30-01	0.01	MRL
	10-01-01–12-31-01	0.011	LRL
Molybdenum (µg/L)	01-01-96–09-26-00	1	MRL
	10-01-00–12-31-01	0.2	LRL
Nickel (µg/L)	01-01-96 09 30 00	1	MRL
	10-01-00–12-31-01	0.06	LRL
Selenium (µg/L)	01-01-96–09-26-99	1	MRL
	10-01-99–09-30-01	2.4	LRL
	10-01-01–12-31-01	2	LRL
Silver (µg/L)	01-01-96–12-31-01	1	MRL
Thallium (µg/L)	01-01-96–11-20-00	0.5	MRL
Uranium (µg/L)	01-01-96–09-26-00	1	MRL
	10-01-00–12-31-01	0.018	LRL
Zinc (µg/L)	01-01-96–12-31-01	1	MRL
Chromium (µg/L)	01-01-96–12-22-97	5	MRL
	12-23-97–09-30-98	14	MRL
	10-01-98–10-17-00	14	LRL
	10-18-00–12-31-01	10	LRL
	01-01-96–09-30-99	1.0	MRL
	10-01-99–05-14-01	0.4	LRL
Hexavalent chromium (µg/L)	01-01-96–12-31-96	1	MRL
1,1-dichlorethene (µg/L)	01-01-96–12-31-01	0.1	MRL
Tetrachloroethene (µg/L)	01-01-96–12-31-01	0.2	MRL
Tetrachloromethane (µg/L)	01-01-96–12-31-01	0.2	MRL
Toluene (µg/L)	01-01-96–12-31-01	0.2	MRL
1,1,1-trichloroethane (µg/L)	01-01-96–12-31-01	0.2	MRL
Trichloroethene (µg/L)	01-01-96–12-31-01	0.2	MRL
Trichloromethane (µg/L)	01-01-96–12-31-01	0.2	MRL
Total organic carbon (mg/L)	01-01-96–09-30-99	0.1	MRL
	10-01-99–10-19-00	0.27	LRL
	10-20-00–12-31-01	0.6	LRL

Table 2. Measured concentrations and normalized absolute differences for gross-alpha radioactivity, gross-beta radioactivity, and cesium-137 from replicate pairs collected from selected sites at the Idaho National Laboratory and vicinity, Idaho, 1996–2001.

[Locations of sites are shown in figures 1–4. Uncertainties are 1σ combined standard uncertainties. Concentrations in table are rounded results, but normalized absolute differences (NADs) were calculated using unrounded concentrations. **Abbreviations:** pCi/L, picocuries per liter, Pu-239, plutonium-239; Cs-137, cesium-137; –, no data]

Site name	Sample collection date	Gross-alpha radioactivity (pCi/L as Pu-239)	(NAD)	Gross-beta radioactivity (pCi/L as Cs-137)	(NAD)	Cesium-137 (pCi/L)	(NAD)
USGS 125	01-09-96	1.4±0.9	0.00	6±2	0.71	8±17	0.84
		1.4±0.9		4±2		30±20	
USGS 87	01-16-96	–	–	–	–	20±20	1.06
		–		–		-10±20	
USGS 120	01-17-96	–	–	–	–	-20±40	0.45
		–		–		0±20	
USGS 83	04-01-96	1.7±0.9	0.22	2±2	1.06	40±30	0.71
		2±1		5±2		10±30	
USGS 34	04-02-96	1.7±0.9	0.83	10±2	1.11	-15±26	0.10
		0.7±0.8		14±3		-20±40	
USGS 5	04-10-96	2±1	0.46	10±2	1.11	-17±25	0.53
		1.4±0.9		14±3		0±20	
USGS 117	04-16-96	–	–	–	–	-60±40	1.20
		–		–		0±30	
USGS 9	04-16-96	1.4±0.9	0.91	3±2	0.35	30±20	0.23
		0.3±0.8		2±2		24±17	
USGS 109	04-17-96	0.7±0.8	0.23	5±2	0.71	0±40	0.33
		1±1		3±2		15±22	
TRA Disposal	04-18-96	–	–	–	–	20±20	0.00
		–		–		20±40	
Site 14	04-24-96	2.4±1.1	0.20	6±2	1.41	30±40	0.37
		2±1		2±2		13±23	
Leo Rogers 1	07-17-96	1±1	0.00	5±2	1.06	13±36	0.06
		1±1		2±2		10±30	
Big Lost River near Arco (BLR nr Arco)	10-07-96	0.3±0.8	1.33	2±2	0.31	20±30	0.47
		2±1		1.1±2.1		0±30	
No Name 1	10-14-96	2±1	1.02	54±5	9.10	10±20	1.11
		0.7±0.8		5±2		-30±30	
USGS 27	10-15-96	1±1	0.74	6±2	0.35	0±20	0.55
		2±1		7±2		-20±30	
USGS 77	10-17-96	1.7±0.9	0.49	8±2	0.71	-10±30	0.16
		2.4±1.1		6±2		-16±24	
USGS 38	10-25-96	1±1	0.55	3±2	8.73	-40±40	1.24
		0.3±0.8		50±5		30±40	
USGS 43	10-28-96	–	–	–	–	-30±30	0.00
		–		–		-30±40	
USGS 60	04-02-97	–	–	–	–	-30±30	0.00
		–		–		-30±20	
USGS 109	04-03-97	1±1	0.30	3±2	0.35	-12±25	0.57
		1.4±0.9		2±2		12±34	
USGS 86	04-03-97	0.7±0.8	0.35	4±2	0.35	-30±30	0.28
		0.3±0.8		3±2		-40±20	
NPR Test	04-08-97	0.3±0.8	0.35	3±2	0.00	20±30	0.00
		0.7±0.8		3±2		20±30	

Table 2.   Measured concentrations and normalized absolute differences for gross-alpha radioactivity, gross-beta radioactivity, and cesium-137 from replicate pairs collected from selected sites at the Idaho National Laboratory and vicinity, Idaho, 1996–2001.—Continued

[Locations of sites are shown in figures 1–4. Uncertainties are 1σ combined standard uncertainties. Concentrations in table are rounded results, but normalized absolute differences (NADs) were calculated using unrounded concentrations. **Abbreviations:** pCi/L, picocuries per liter, Pu-239, plutonium-239; Cs-137, cesium-137; –, no data]

Site name	Sample collection date	Gross-alpha radioactivity (pCi/L as Pu-239)	(NAD)	Gross-beta radioactivity (pCi/L as Cs-137)	(NAD)	Cesium-137 (pCi/L)	(NAD)
Big Lost River at Experimental Dairy Farm near Howe (BLR at exp dairy farm nr Howe)	04-09-97	1±1 1±1	0.00	2±2 2±2	0.00	-70±30 -20±20	1.39
USGS 63	04-22-97	– –	–	– –	–	1±16 -11±24	0.42
USGS 58	05-05-97	– –	–	– –	–	-14±22 -30±20	0.54
USGS 107	10-01-97	2.8±1.2 1.4±0.9	0.93	3±2 2±2	0.35	0±30 30±20	0.83
USGS 108	10-07-97	2.4±1.1 1.7±0.9	0.49	1.3±2.1 5±2	1.28	-10±21 20±20	1.03
USGS 37	10-08-97	– –	–	– –	–	0±30 13±29	0.31
USGS 8	10-08-97	2±1 2±1	0.00	1.1±2.1 0±2	0.38	-30±30 70±40	2.00
USGS 101	10-16-97	2±1 1±1	0.71	4±2 3±2	0.35	0±20 0±20	0.00
ANP 9	10-27-97	0.7±0.8 2±1	1.02	3±2 3±2	0.00	-12±24 -20±40	0.17
USGS 5	03-31-98	0.7±0.8 1.7±0.9	0.83	4±2 3±2	0.35	40±40 12±36	0.52
USGS 117	04-01-98	– –	–	– –	–	-20±30 -20±30	0.00
USGS 14	04-07-98	1±1 1±1	0.00	3±2 4±2	0.35	40±40 40±40	0.00
USGS 19	04-08-98	0.7±0.8 1.4±0.9	0.58	4±2 1.7±0.8	1.07	13±36 -14±25	0.62
USGS 7	04-13-98	1±1 1.7±0.9	0.52	6±2 2.9±0.9	1.41	20±40 -50±40	1.24
Highway 3	04-20-98	1.4±0.9 0.7±0.8	0.58	3±2 3.3±0.9	0.14	0±40 -11±35	0.21
USGS 65	04-28-98	1.4±0.9 2.1±1.0	0.52	4±2 6±2	0.71	30±40 -10±30	0.80
USGS 90	07-21-98	– –	–	– –	–	10±20 4±40	0.13
USGS 97	10-14-98	1.7±0.9 2.4±1.1	0.49	3±2 3±2	0.00	20±20 19±40	0.02
USGS 1	10-20-98	0.7±0.8 0.7±0.8	0.00	3±2 4±2	0.35	-30±40 50±40	1.41
USGS 110A	10-20-98	0.7±0.8 0.3±0.8	0.35	4±2 5±2	0.35	40±20 3±16	1.44
USGS 76	10-22-98	– –	–	– –	–	-30±20 0±30	0.83
USGS 47	10-28-98	– –	–	– –	–	20±40 -10±20	0.67

Table 2. Measured concentrations and normalized absolute differences for gross-alpha radioactivity, gross-beta radioactivity, and cesium-137 from replicate pairs collected from selected sites at the Idaho National Laboratory and vicinity, Idaho, 1996–2001.—Continued

[Locations of sites are shown in figures 1–4. Uncertainties are 1σ combined standard uncertainties. Concentrations in table are rounded results, but normalized absolute differences (NADs) were calculated using unrounded concentrations. **Abbreviations:** pCi/L, picocuries per liter, Pu-239, plutonium-239; Cs-137, cesium-137; –, no data]

Site name	Sample collection date	Gross-alpha radioactivity (pCi/L as Pu-239)	(NAD)	Gross-beta radioactivity (pCi/L as Cs-137)	(NAD)	Cesium-137 (pCi/L)	(NAD)
USGS 88	01-13-99	–	–	–	–	16±40	0.56
		–		–		-12±30	
USGS 40	01-19-99	–	–	–	–	39±20	0.69
		–		–		8±40	
USGS 83	04-01-99	0.7±0.8	0.25	3±2	0.00	-11±40	1.14
		1±1		3±2		46±30	
USGS 12	04-05-99	1.4±0.9	0.24	2±2	0.18	-2±30	0.24
		1.7±0.9		1±2		10±40	
EBR 1	04-13-99	1.7±0.9	0.55	3±2	0.00	8±20	1.03
		1±1		3±2		-21±20	
USGS 61	04-13-99	–	–	–	–	10±20	0.71
		–		–		-10±20	
P AND W 2	04-14-99	1±1	0.55	2±2	0.00	-19±40	0.24
		0.3±0.8		2±2		-7±20	
Big Lost River below INL Diversion near Arco (BLR blw INL div nr Arco)	04-20-99	1±0.9	0.25	2±2	0.35	10±30	0.26
		0.7±0.8		3±2		21±30	
PW 2	04-22-99	–	–	–	–	30±30	1.69
		–		–		-31±20	
USGS 46	04-22-99	–	–	–	–	49±30	0.61
		–		–		23±30	
CPP 1	04-26-99	0.3±0.8	1.16	5±2	0.35	25±10	1.74
		1.7±0.9		6±2		-14±20	
USGS 11	10-07-99	1.0±0.8	0.70	2±2	0.71	30±30	0.28
		0.3±0.6		4±2		20±20	
Big Lost River below Mackay Reservoir (BLR blw Mackay Reservoir)	10-07-99	2±1	0.71	1±2	0.35	-30±20	0.00
		1±1		2±2		-30±40	
USGS 34	10-14-99	0.7±0.7	0.28	10±2	0.00	-10±20	0.28
		1.0±0.8		10±3		-20±30	
USGS 50	10-20-99	–	–	–	–	20±20	1.05
		–		–		-17±29	
USGS 58	10-21-99	–	–	–	–	-30±30	0.55
		–		–		-10±20	
USGS 105	10-25-99	1.0±0.8	0.00	4±2	0.35	52±33	1.02
		1.0±0.9		5±2		13±19	
USGS 23	10-27-99	2.0±0.9	0.79	6±2	1.06	12±31	0.74
		1.0±0.9		3±2		-20±30	
PSTF Test	04-04-00	2±1	0.71	5±2	0.00	-16±23	0.11
		1±1		5±2		-20±30	
USGS 98	04-04-00	1±1	0.65	3±2	0.71	28±16	0.53
		0.3±0.6		5±2		10±30	
USGS 12	04-05-00	0.7±0.7	0.28	3±2	0.35	10±20	1.11
		1±1		4±2		-30±30	
USGS 119	04-06-00	–	–	–	–	30±30	0.83
		–		–		0±20	
USGS 125	04-11-00	1±0.8	0.00	3±2	0.35	0±20	0.00
		1±0.8		4±2		0±20	

Table 2. Measured concentrations and normalized absolute differences for gross-alpha radioactivity, gross-beta radioactivity, and cesium-137 from replicate pairs collected from selected sites at the Idaho National Laboratory and vicinity, Idaho, 1996–2001.—Continued

[Locations of sites are shown in figures 1–4. Uncertainties are 1σ combined standard uncertainties. Concentrations in table are rounded results, but normalized absolute differences (NADs) were calculated using unrounded concentrations. **Abbreviations:** pCi/L, picocuries per liter, Pu-239, plutonium-239; Cs-137, cesium-137; –, no data]

Site name	Sample collection date	Gross-alpha radioactivity (pCi/L as Pu-239)	(NAD)	Gross-beta radioactivity (pCi/L as Cs-137)	(NAD)	Cesium-137 (pCi/L)	(NAD)
USGS 26	04-12-00	1±1	0.00	3±2	0.71	-50±30	2.22
		1±1		5±2		30±20	
USGS 27	04-12-00	2±1	1.07	3±2	0.35	10±20	0.35
		0.7±0.7		4±2		0±20	
USGS 97	09-27-00	1±1	0.92	3±2	0.00	10±30	0.00
		0.0±0.6		3±2		10±30	
USGS 58	10-04-00	–	–	–	–	10±30	0.71
		–		–		40±30	
USGS 11	10-05-00	1±1	0.70	3±2	0.35	-20±20	0.30
		0.3±0.6		2±2		-11±22	
USGS 57	10-05-00	–	–	–	–	20±40	0.80
		–		–		60±30	
USGS 77	10-06-00	0.7±0.7	0.43	10±3	0.00	11±25	0.40
		0.3±0.6		10±3		30±40	
USGS 50	10-10-00	–	–	–	–	10±30	0.00
		–		–		10±30	
RWMC M13S	10-17-00	0.7±0.7	0.00	4±2	0.00	0±30	0.00
		0.7±0.7		4±2		0±40	
USGS 88	01-23-01	–	–	–	–	30±30	0.28
		–		–		20±20	
USGS 73	04-02-01	–	–	–	–	56±39	0.82
		–		–		20±20	
USGS 103	04-04-01	0.0±0.6	1.00	1.4±2.1	0.35	-11±35	0.40
		1±1		2.5±2.2		10±40	
USGS 17	04-04-01	0.3±0.6	0.70	1.5±2.1	1.16	-10±20	0.83
		1±1		5.2±2.3		20±30	
Site 14	04-05-01	0.7±0.7	1.07	1±2	0.35	50±33	1.30
		2±1		2±2		0±20	
USGS 9	04-11-01	1±1	0.28	3.2±2.2	0.74	0±30	0.47
		0.7±0.7		1±2		-20±30	
PW 4	04-12-01	–	–	–	–	20±30	0.00
		–		–		20±16	
RWMC M14S	04-17-01	0.3±0.6	1.02	2±2	0.24	-30±20	1.11
		1.4±0.9		2.7±2.2		10±30	
USGS 127	04-18-01	1.4±0.9	0.61	2±2	0.47	-10±30	0.76
		0.7±0.7		3.4±2.2		15±14	
USGS 44	04-24-01	–	–	–	–	30±30	0.42
		–		–		15±20	
CFA LF 2-10	04-30-01	0.0±0.6	0.35	5.1±2.3	0.50	15±30	0.30
		0.3±0.6		3.5±2.2		0±40	
USGS 84	04-30-01	1±1	0.00	-1.6±2.0	1.00	-20±40	0.16
		1±1		1.3±2.1		-30±50	
USGS 126A	07-10-01	1±1	0.33	4±2	0.95	30±30	0.60
		1.4±0.9		1.3±2.0		0±40	

Table 2. Measured concentrations and normalized absolute differences for gross-alpha radioactivity, gross-beta radioactivity, and cesium-137 from replicate pairs collected from selected sites at the Idaho National Laboratory and vicinity, Idaho, 1996–2001.—Continued

[Locations of sites are shown in figures 1–4. Uncertainties are 1σ combined standard uncertainties. Concentrations in table are rounded results, but normalized absolute differences (NADs) were calculated using unrounded concentrations. **Abbreviations:** pCi/L, picocuries per liter, Pu-239, plutonium-239; Cs-137, cesium-137; –, no data]

Site name	Sample collection date	Gross-alpha radioactivity		Gross-beta radioactivity		Cesium-137	
		(pCi/L as Pu-239)	(NAD)	(pCi/L as Cs-137)	(NAD)	(pCi/L)	(NAD)
USGS 126B	07-10-01	0.3±0.6	0.35	1.1±2.0	1.03	-20±30	1.65
		0.0±0.6		4±2		50±30	
USGS 120	07-12-01	–	–	–	–	10±30	0.20
		–		–		20±40	
USGS 38	10-11-01	0.7±0.7	0.28	45±4	1.87	40±20	0.28
		1±1		57±5		30±30	
USGS 43	10-22-01	–	–	–	–	20±30	0.07
		–		–		17±34	

Table 3. Measured concentrations and normalized absolute differences for tritium and strontium-90 from replicate pairs collected from selected sites at the Idaho National Laboratory and vicinity, Idaho, 1996–2001.

[Locations of sites are shown in figures 1–4. Uncertainties are 1σ combined standard uncertainties. **Abbreviations:** pCi/L, picocuries per liter; NAD, normalized absolute difference; –, no data]

Site name	Sample collection date	Tritium (pCi/L)	Tritium (NAD)	Strontium-90 (pCi/L)	Strontium-90 (NAD)
USGS 103	01-02-96	50±80	1.15	–	–
		180±80		–	
USGS 125	01-09-96	190±80	0.09	–	–
		180±80		–	
RWMC Production	01-16-96	1,500±200	1.41	0±0.8	0.78
		1,900±200		-1.4±1.6	
USGS 87	01-16-96	1,400±200	0.71	-2.5±1.2	1.45
		1,600±200		0.4±1.6	
USGS 120	01-17-96	300±200	0.00	-1.8±1.2	0.55
		300±200		-0.7±1.6	
USGS 112	01-18-96	14,800±700	0.71	21.2±1.3	6.79
		15,500±700		-1±3	
USGS 83	04-01-96	-90±90	0.47	–	–
		-150±90		–	
USGS 34	04-02-96	3,400±300	0.94	1.5±0.9	0.00
		3,800±300		1.5±0.9	
USGS 79	04-02-96	-100±200	0.71	–	–
		100±200		–	
CFA 2	04-03-96	14,700±700	0.20	0±3	0.38
		14,500±700		-1.2±0.9	
Atomic City	04-04-96	-90±90	0.05	–	–
		-100±200		–	
USGS 5	04-10-96	-120±190	0.07	–	–
		-100±200		–	
USGS 20	04-11-96	7,100±400	0.53	0±0.7	0.00
		7,400±400		0±0.7	
USGS 117	04-16-96	0±200	0.00	0.6±0.7	1.62
		0±200		-1±1	
USGS 9	04-16-96	-50±90	0.71	–	–
		40±90		–	
MTR Test	04-17-96	2,000±300	0.47	–	–
		2,200±300		–	
USGS 109	04-17-96	70±90	0.00	–	–
		70±90		–	
TRA Disposal	04-18-96	6,800±400	0.35	0.5±0.7	1.13
		6,600±400		-0.7±0.8	
USGS 52	04-22-96	6,000±400	0.18	11±1	1.68
		6,100±400		13.5±1.1	
SITE 14	04-24-96	0±200	0.40	–	–
		-110±190		–	

Table 3. Measured concentrations and normalized absolute differences for tritium and strontium-90 from replicate pairs collected from selected sites at the Idaho National Laboratory and vicinity, Idaho, 1996–2001.—Continued

[Locations of sites are shown in figures 1–4. Uncertainties are 1σ combined standard uncertainties. **Abbreviations:** pCi/L, picocuries per liter; NAD, normalized absolute difference; –, no data]

Site name	Sample collection date	Tritium		Strontium-90	
		(pCi/L)	(NAD)	(pCi/L)	(NAD)
Leo Rogers 1	07-17-96	160±80	0.09	–	–
		150±80		–	
IET 1 Disposal	07-18-96	300±200	0.71	1.1±0.7	0.91
		500±200		0.2±0.7	
TRA 1	07-18-96	100±200	0.71	–	–
		-100±200		–	
TRA 3	07-18-96	300±200	0.71	–	–
		100±200		–	
ANP 6	07-19-96	200±200	1.41	0.5±0.7	0.57
		600±200		0±1	
RWMC M3S	07-22-96	2,000±300	0.47	-1±1	1.15
		2,200±300		0.3±0.8	
USGS 66	07-25-96	5,100±400	0.53	0.6±0.8	0.27
		5,400±400		0.9±0.8	
USGS 72	07-30-96	0±200	0.35	0.12±0.75	0.98
		100±200		1.2±0.8	
Big Lost River near Arco (BLR nr Arco)	10-07-96	-300±200	0.71	–	–
		-500±200		–	
Birch Creek at Blue Dome Inn near Reno (BC at Blue Dome nr Reno)	10-08-96	-300±200	1.77	–	–
		200±200		–	
No Name 1	10-14-96	-150±220	0.50	-0.7±0.7	1.21
		-300±200		0.5±0.7	
USGS 27	10-15-96	-150±220	0.84	–	–
		-400±200		–	
SPERT 1	10-16-96	150±230	0.85	–	–
		-120±220		–	
USGS 77	10-17-96	24,000±1000	0.14	1.4±0.7	0.91
		24,200±1000		0.5±0.7	
USGS 48	10-21-96	10,300±500	0.28	31±2	0.00
		10,100±500		31±2	
PW 3	10-23-96	140±220	0.20	1.4±0.8	0.35
		200±200		1±1	
USGS 73	10-23-96	78,200±2700	0.29	0.3±0.7	0.40
		77,100±2700		0.7±0.7	
USGS 38	10-25-96	11,800±600	0.12	26±1	0.05
		11,900±600		26.1±1.4	
USGS 43	10-28-96	4,400±400	0.35	1.3±0.7	0.19
		4,600±400		1.5±0.8	
USGS 113	01-09-97	9,200±400	0.18	12.6±1.1	0.64
		9,300±400		13.6±1.1	

Table 3.   Measured concentrations and normalized absolute differences for tritium and strontium-90 from replicate pairs collected from selected sites at the Idaho National Laboratory and vicinity, Idaho, 1996–2001.—Continued

[Locations of sites are shown in figures 1–4. Uncertainties are 1σ combined standard uncertainties. **Abbreviations:** pCi/L, picocuries per liter; NAD, normalized absolute difference; –, no data]

Site name	Sample collection date	Tritium		Strontium-90	
		(pCi/L)	(NAD)	(pCi/L)	(NAD)
USGS 54	01-23-97	1,900±200 1,700±200	0.71	119±4 111±4	1.41
USGS 60	04-02-97	410±130 550±130	0.76	6.6±0.9 5.8±0.9	0.63
USGS 109	04-03-97	80±50 20±50	0.85	– –	–
USGS 86	04-03-97	0±50 -20±50	0.28	– –	–
CPP 5	04-08-97	0±110 -80±100	0.54	0.2±0.7 -0.5±0.7	0.71
NPR Test	04-08-97	-170±100 -120±100	0.35	– –	–
USGS 41	04-08-97	1,700±200 1,700±200	0.00	13±1 13±1	0.00
USGS 42	04-08-97	4,500±300 4,900±300	0.94	21.7±1.4 19±1	1.41
Big Lost River at Experimental Dairy Farm near Howe (BLR at exp dairy farm nr Howe)	04-09-97	-150±100 10±110	1.08	– –	–
USGS 124	04-10-97	60±50 120±50	0.85	– –	–
USGS 63	04-22-97	530±130 620±130	0.49	4.3±0.8 4.1±0.9	0.17
USGS 104	04-29-97	1,650±90 1,570±90	0.63	– –	–
USGS 58	05-05-97	3,900±300 3,500±200	1.11	-0.4±0.7 -1.1±0.7	0.71
USGS 6	07-01-97	-110±110 -60±110	0.32	-0.6±0.7 1.1±0.7	1.72
USGS 82	07-02-97	580±140 960±150	1.85	0.1±0.7 0.8±0.7	0.73
USGS 32	07-08-97	20±110 -20±110	0.26	-0.3±0.7 1.4±0.7	1.72
CWP 1	07-15-97	130±120 -30±110	0.98	-0.2±0.7 0.8±0.7	1.01
Badging Facility	07-22-97	-10±110 -30±110	0.13	0.3±0.7 -0.1±0.7	0.40
USGS 107	10-01-97	-40±40 20±50	0.94	– –	–
USGS 108	10-07-97	110±50 110±50	0.00	– –	–

Table 3. Measured concentrations and normalized absolute differences for tritium and strontium-90 from replicate pairs collected from selected sites at the Idaho National Laboratory and vicinity, Idaho, 1996–2001.—Continued

[Locations of sites are shown in figures 1–4. Uncertainties are 1σ combined standard uncertainties. **Abbreviations:** pCi/L, picocuries per liter; NAD, normalized absolute difference; –, no data]

Site name	Sample collection date	Tritium (pCi/L)	Tritium (NAD)	Strontium-90 (pCi/L)	Strontium-90 (NAD)
USGS 39	10-07-97	2,500±200 3,100±200	2.12	-0.6±0.5 -0.1±0.8	0.53
USGS 37	10-08-97	10,200±500 11,600±500	1.98	9.9±1.2 8.9±1.2	0.65
USGS 8	10-08-97	30±50 50±50	0.28	– –	–
USGS 101	10-16-97	40±110 -50±100	0.61	– –	–
PW 8	10-21-97	580±130 690±140	0.58	19±1 19.5±1.3	0.30
SITE 4	10-21-97	40±110 10±110	0.19	– –	–
USGS 67	10-21-97	11,800±500 13,700±600	2.43	16±1.5 17±1.5	0.47
USGS 115	10-22-97	3,100±200 3,800±300	1.94	-0.1±0.5 0.6±0.5	0.99
ANP 9	10-27-97	180±120 70±110	0.68	-1.2±1.2 0.7±0.7	1.37
USGS 51	10-27-97	16,300±700 17,600±700	1.31	-0.5±0.6 0.8±0.8	1.30
USGS 119	01-13-98	-180±110 -140±110	0.26	0±1 0.6±0.5	0.77
USGS 104	01-15-98	1,520±90 1,690±90	1.34	– –	–
USGS 114	01-22-98	16,800±700 20,100±800	3.10	1.3±0.5 1.3±0.5	0.00
PW 1	01-26-98	0±120 40±120	0.24	0.5±0.6 1.3±0.5	1.02
USGS 5	03-31-98	-100±110 -50±110	0.32	– –	–
USGS 117	04-01-98	-60±110 -110±110	0.32	0.9±1.6 -0.6±0.6	0.88
USGS 121	04-02-98	-80±110 20±110	0.64	-0.6±0.6 -0.7±0.6	0.12
USGS 106	04-06-98	1,380±90 1,400±90	0.16	– –	–
USGS 111	04-06-98	7,000±400 7,100±400	0.18	-0.2±0.6 -1.1±0.7	0.98
USGS 14	04-07-98	-30±50 -60±50	0.42	– –	–

Table 3. Measured concentrations and normalized absolute differences for tritium and strontium-90 from replicate pairs collected from selected sites at the Idaho National Laboratory and vicinity, Idaho, 1996–2001.— Continued

[Locations of sites are shown in figures 1–4. Uncertainties are 1σ combined standard uncertainties. **Abbreviations:** pCi/L, picocuries per liter; NAD, normalized absolute difference; –, no data]

Site name	Sample collection date	Tritium (pCi/L)	Tritium (NAD)	Strontium-90 (pCi/L)	Strontium-90 (NAD)
USGS 19	04-08-98	-150±110	0.39	–	–
		-90±110		–	
USGS 7	04-13-98	10±110	0.71	0.9±0.6	0.71
		-100±110		0.3±0.6	
Highway 3	04-20-98	-40±110	0.86	–	–
		100±120		–	
Arbor Test	04-22-98	10±110	0.51	–	–
		-70±110		–	
USGS 65	04-28-98	17,600±700	0.61	0.2±0.4	2.71
		17,000±700		-8±3	
USGS 85	04-28-98	6,000±300	0.71	2.1±0.6	1.71
		5,700±300		-0.5±1.4	
USGS 69	07-01-98	50±110	1.14	0±1	0.54
		-120±100		-0.5±0.6	
RWMC M3S	07-07-98	1,600±200	0.00	-0.2±0.6	0.35
		1,600±200		0.1±0.6	
USGS 18	07-20-98	10±110	0.00	0.5±0.6	0.94
		10±110		1.3±0.6	
TRA 4	07-21-98	-70±110	0.77	–	–
		50±110		–	
USGS 90	07-21-98	1,380±170	0.25	0.5±0.7	0.75
		1,440±170		1.3±0.8	
USGS 2	07-22-98	10±110	0.06	1.2±0.6	0.24
		20±110		1±1	
USGS 36	07-22-98	6,500±300	2.83	6.5±0.6	7.70
		5,300±300		13.6±0.7	
CPP 2	09-30-98	30±110	0.13	0.4±0.5	0.71
		10±110		-0.1±0.5	
Mud Lake near Terreton (ML nr Terreton)	10-06-98	160±120	1.23	–	–
		-40±110		–	
USGS 97	10-14-98	110±110	0.43	-1±1	1.50
		180±120		0.7±0.8	
USGS 35	10-15-98	1,900±200	0.35	1.6±0.6	0.47
		1,800±200		1.2±0.6	
USGS 1	10-20-98	40±50	1.27	–	–
		-50±50		–	
USGS 110A	10-20-98	10±50	0.33	–	–
		-13±48		–	
USGS 123	10-20-98	13,900±600	0.12	30.9±1.3	0.21
		14,000±600		30.5±1.4	

Table 3. Measured concentrations and normalized absolute differences for tritium and strontium-90 from replicate pairs collected from selected sites at the Idaho National Laboratory and vicinity, Idaho, 1996–2001.—Continued

[Locations of sites are shown in figures 1–4. Uncertainties are 1σ combined standard uncertainties. **Abbreviations:** pCi/L, picocuries per liter; NAD, normalized absolute difference; –, no data]

Site name	Sample collection date	Tritium		Strontium-90	
		(pCi/L)	(NAD)	(pCi/L)	(NAD)
WS INEL 1	10-20-98	80±110	0.51	–	–
		0±110		–	
USGS 76	10-22-98	2,000±200	0.00	1.4±0.8	0.80
		2,000±200		0.5±0.8	
PW 6	10-27-98	9,700±500	0.57	0±1	0.59
		10,100±500		0.5±0.6	
USGS 47	10-28-98	4,600±300	0.47	41±2	1.32
		4,800±300		43.9±1.6	
USGS 68	12-01-98	540±130	2.94	0.7±0.5	0.57
		40±110		1.1±0.5	
USGS 88	01-13-99	0±120	0.59	1.1±0.8	0.00
		-100±120		1.1±0.8	
USGS 40	01-19-99	2,200±200	0.71	15±1	0.00
		2,000±200		15±1	
USGS 99	03-31-99	30±120	1.66	–	–
		-230±100		–	
USGS 79	04-01-99	770±150	0.78	–	–
		610±140		–	
USGS 83	04-01-99	-50±50	0.86	–	–
		11±50		–	
USGS 12	04-05-99	-30±110	0.49	–	–
		50±120		–	
EBR 1	04-13-99	-40±110	0.32	–	–
		10±110		–	
USGS 61	04-13-99	8,100±400	0.53	0.5±0.8	0.97
		8,400±400		1.6±0.8	
P AND W 2	04-14-99	-60±110	0.39	–	–
		0±110		–	
USGS 100	04-19-99	-110±110	0.19	–	–
		-140±110		–	
Big Lost River below INL Diversion near Arco (BLR blw INL div nr Arco)	04-20-99	120±120	1.11	–	–
		-60±110		–	
PW 2	04-22-99	130±120	0.06	2±1	0.00
		140±120		2±1	
USGS 46	04-22-99	760±150	0.19	9.1±0.8	0.62
		800±150		9.8±0.8	
CPP 1	04-26-99	90±120	0.29	0.1±0.7	0.40
		40±120		0.5±0.7	
CWP 3	06-30-99	10±120	0.12	1.8±0.7	0.75
		30±120		1.0±0.8	

Table 3. Measured concentrations and normalized absolute differences for tritium and strontium-90 from replicate pairs collected from selected sites at the Idaho National Laboratory and vicinity, Idaho, 1996–2001.— Continued

[Locations of sites are shown in figures 1–4. Uncertainties are 1σ combined standard uncertainties. **Abbreviations:** pCi/L, picocuries per liter; NAD, normalized absolute difference; –, no data]

Site name	Sample collection date	Tritium (pCi/L)	Tritium (NAD)	Strontium-90 (pCi/L)	Strontium-90 (NAD)
USGS 22	07-07-99	50±120	0.41	–	–
		120±120		–	
SPERT 1	07-08-99	-50±110	0.80	–	–
		80±120		–	
RWMC M7S	07-12-99	1,400±200	0.00	1.2±0.7	0.56
		1,400±200		0.6±0.8	
USGS 15	07-13-99	10±120	0.41	-0.3±0.8	0.47
		80±120		0.2±0.7	
TRA 1	07-21-99	-50±110	0.13	–	–
		-30±110		–	
PW 4	10-06-99	480±150	0.09	3.4±0.8	0.35
		460±150		3.8±0.8	
USGS 114	10-06-99	16,600±700	1.01	0.6±0.7	0.00
		17,600±700		0.6±0.8	
USGS 11	10-07-99	-40±50	0.28	–	–
		-20±50		–	
Big Lost River below Mackay Reservoir (BLR blw MacKay Reservoir)	10-07-99	-100±130	0.22	–	–
		-60±130		–	
USGS 20	10-14-99	7,600±400	0.88	-0.1±0.8	0.37
		7,100±400		0.3±0.7	
USGS 34	10-14-99	1,500±200	0.35	3.1±0.8	0.27
		1,400±200		3.4±0.8	
CFA 1	10-20-99	13,900±600	0.12	0.4±0.8	0.28
		14,000±600		0.7±0.7	
USGS 50	10-20-99	35,200±1300	0.33	204±5	0.85
		34,600±1300		210±5	
USGS 58	10-21-99	1,600±200	0.00	0.8±0.8	0.19
		1,600±200		1±1	
PW 5	10-25-99	-50±130	0.30	2.3±0.8	0.18
		0±100		2.1±0.8	
USGS 105	10-25-99	150±60	0.64	–	–
		100±50		–	
USGS 121	10-25-99	40±130	0.21	-0.5±0.8	1.41
		80±140		1.1±0.8	
USGS 52	10-25-99	2,300±200	0.00	6.5±0.8	0.09
		2,300±200		6.6±0.8	
USGS 23	10-27-99	-130±130	0.76	–	–
		10±130		–	
RWMC Production	01-13-00	1,330±170	0.17	0.79±0.73	0.04
		1,290±170		0.84±0.74	

Table 3.   Measured concentrations and normalized absolute differences for tritium and strontium-90 from replicate pairs collected from selected sites at the Idaho National Laboratory and vicinity, Idaho, 1996–2001.— Continued

[Locations of sites are shown in figures 1–4. Uncertainties are 1σ combined standard uncertainties. **Abbreviations:** pCi/L, picocuries per liter; NAD, normalized absolute difference; –, no data]

Site name	Sample collection date	Tritium		Strontium-90	
		(pCi/L)	(NAD)	(pCi/L)	(NAD)
CFA 2	01-19-00	9,700±500	0.28	–	–
		9,900±500		–	
USGS 89	01-20-00	-30±110	0.74	1.1±0.8	0.47
		-140±100		0.6±0.7	
PSTF Test	04-04-00	-140±110	0.26	1±1	0.55
		-180±110		0.3±0.8	
USGS 98	04-04-00	-80±120	0.55	0.7±0.7	0.35
		-170±110		1.1±0.9	
USGS 12	04-05-00	-80±120	0.06	–	–
		-90±120		–	
USGS 119	04-06-00	-170±110	0.13	1.1±0.7	0.00
		-150±110		1.1±0.8	
MTR Test	04-10-00	-120±110	0.49	–	–
		-40±120		–	
Atomic City	04-11-00	-50±50	0.66	–	–
		-130±110		–	
Little Lost River near Howe (LLR nr Howe)	04-11-00	-140±110	0.06	–	–
		-150±110		–	
USGS 125	04-11-00	-100±110	0.06	–	–
		-90±120		–	
USGS 26	04-12-00	-210±110	0.26	2±1	0.23
		-170±110		1.8±0.8	
USGS 27	04-12-00	-170±110	0.06	–	–
		-160±110		–	
USGS 102	07-10-00	-60±110	0.86	1.7±0.7	0.94
		80±120		0.7±0.8	
USGS 31	07-10-00	0±120	0.35	1.3±0.7	0.66
		60±120		0.6±0.8	
Badging Facility	07-13-00	-10±120	0.29	2±1	1.32
		40±120		0.6±0.7	
RWMC Production	07-13-00	1,500±200	0.35	0.8±0.7	0.09
		1,400±200		0.7±0.8	
SITE 19	07-13-00	620±150	1.21	–	–
		380±130		–	
CWP 2	07-17-00	30±120	0.12	0.4±0.7	0.40
		10±120		0±1	
Area 2	07-18-00	-60±110	0.37	1.8±0.8	0.71
		0±120		1±1	
USGS 66	07-25-00	2,600±200	0.00	0.7±0.9	0.33
		2,600±200		0.3±0.8	

Table 3.   Measured concentrations and normalized absolute differences for tritium and strontium-90 from replicate pairs collected from selected sites at the Idaho National Laboratory and vicinity, Idaho, 1996–2001.—Continued

[Locations of sites are shown in figures 1–4. Uncertainties are 1σ combined standard uncertainties. **Abbreviations:** pCi/L, picocuries per liter; NAD, normalized absolute difference; –, no data]

Site name	Sample collection date	Tritium (pCi/L)	Tritium (NAD)	Strontium-90 (pCi/L)	Strontium-90 (NAD)
USGS 62	09-27-00	160±130 50±120	0.62	2.9±0.9 3.3±0.9	0.31
USGS 97	09-27-00	-150±110 -130±120	0.12	1.3±0.7 -0.1±0.7	1.41
USGS 82	09-28-00	-210±110 0±120	1.29	0.4±0.9 1.1±0.9	0.55
USGS 58	10-04-00	1,200±200 1,050±170	0.57	-0.7±0.7 0.7±0.7	1.41
USGS 60	10-04-00	120±130 130±130	0.05	8±1 7±1	0.71
USGS 11	10-05-00	11.6±53.1 -13.9±52.5	0.34	– –	–
USGS 124	10-05-00	180±60 210±60	0.35	– –	–
USGS 57	10-05-00	5,600±300 6,400±400	1.60	15.7±0.9 17±1	0.74
USGS 77	10-06-00	11,800±500 13,300±600	1.92	1.6±0.7 2±1	0.40
USGS 50	10-10-00	28,000±1100 27,900±1100	0.06	172±4 168±4	0.71
Birch Creek at Blue Dome Inn near Reno (BC at Blue Dome nr Reno)	10-17-00	-180±110 -240±110	0.39	– –	–
RWMC M13S	10-17-00	-280±110 -140±120	0.86	– –	–
USGS 104	10-23-00	1,050±170 1,020±170	0.12	– –	–
Atomic City	10-24-00	-80±50 -40±50	0.57	– –	–
USGS 116	01-10-01	2,900±200 2,800±200	0.35	1±1 0.8±0.6	0.24
USGS 88	01-23-01	40±130 -200±110	1.41	0.2±0.7 1.3±0.7	1.11
USGS 73	04-02-01	19,900±600 20,800±600	1.06	0.9±0.7 0.7±0.7	0.22
USGS 103	04-04-01	-140±50 -80±60	0.77	– –	–
USGS 17	04-04-01	-80±130 2,000±200	8.72	– –	–
SITE 14	04-05-01	-60±130 -60±130	0.00	– –	–

Table 3. Measured concentrations and normalized absolute differences for tritium and strontium-90 from replicate pairs collected from selected sites at the Idaho National Laboratory and vicinity, Idaho, 1996–2001.—Continued

[Locations of sites are shown in figures 1–4. Uncertainties are 1σ combined standard uncertainties. **Abbreviations:** pCi/L, picocuries per liter; NAD, normalized absolute difference; –, no data]

Site name	Sample collection date	Tritium		Strontium-90	
		(pCi/L)	(NAD)	(pCi/L)	(NAD)
Birch Creek at Blue Dome Inn near Reno (BC at Blue Dome nr Reno)	04-11-01	-40±130 -30±130	0.05	– –	–
USGS 9	04-11-01	-50±60 -70±60	0.24	– –	–
PW 4	04-12-01	50±130 110±140	0.31	0.8±0.8 2.3±0.7	1.41
RWMC M14S	04-17-01	1,800±200 1,600±200	0.71	– –	–
USGS 127	04-18-01	60±140 0±130	0.31	0.3±0.7 1.6±0.7	1.31
USGS 41	04-23-01	1,700±200 1,900±200	0.71	12.8±0.7 12.9±0.7	0.10
USGS 44	04-24-01	130±140 50±140	0.40	3.9±0.7 2.9±0.7	1.01
CFA LF 2-10	04-30-01	1,800±200 -110±130	8.01	1.8±0.7 -0.4±0.8	2.07
USGS 84	04-30-01	2,600±200 2,500±200	0.35	0.5±0.6 0.8±0.7	0.33
USGS 45	05-01-01	260±150 310±150	0.24	1.4±0.6 1.3±0.6	0.12
SITE 9	07-05-01	-240±110 -160±120	0.49	-0.3±0.5 0.1±0.6	0.51
USGS 126A	07-10-01	-90±120 -20±120	0.41	– –	–
USGS 126B	07-10-01	-100±120 -70±120	0.18	– –	–
USGS 120	07-12-01	120±130 20±130	0.54	2±1 0.7±0.7	1.31
USGS 29	07-16-01	-60±120 -60±120	0.00	0.1±0.6 0±1	0.13
TRA 3	07-19-01	-150±120 -130±120	0.12	– –	–
SITE 17	07-30-01	-10±120 -110±120	0.58	0.5±0.6 -0.4±0.6	1.06
Birch Creek at Blue Dome Inn near Reno (BC at Blue Dome nr Reno)	10-04-01	0±120 -20±120	0.12	– –	–
PW 8	10-09-01	490±140 450±140	0.20	9.3±0.8 10±0.8	0.62
USGS 38	10-11-01	6,000±400 5,800±400	0.35	17±1 17.8±0.9	0.74

Table 3. Measured concentrations and normalized absolute differences for tritium and strontium-90 from replicate pairs collected from selected sites at the Idaho National Laboratory and vicinity, Idaho, 1996–2001.—Continued

[Locations of sites are shown in figures 1–4. Uncertainties are 1σ combined standard uncertainties. **Abbreviations:** pCi/L, picocuries per liter; NAD, normalized absolute difference; –, no data]

Site name	Sample collection date	Tritium		Strontium-90	
		(pCi/L)	(NAD)	(pCi/L)	(NAD)
Arbor Test	10-15-01	20±120	0.29	–	–
		-30±120		–	
USGS 106	10-17-01	1,000±80	1.41	–	–
		840±80		–	
USGS 42	10-17-01	760±160	0.00	7.2±0.7	0.61
		760±160		6.6±0.7	
USGS 43	10-22-01	2,900±200	1.06	-2.4±0.7	1.79
		2,600±200		-0.5±0.8	

Table 4. Measured concentrations and normalized absolute differences for plutonium-238, plutonium-239+240, and americium-241 from replicate pairs collected from selected sites at the Idaho National Laboratory and vicinity, Idaho, 1996–2001.

[Locations of sites are shown in figures 1–4. Uncertainties are 1σ combined standard uncertainties. **Abbreviations:** pCi/L, picocuries per liter; NAD, normalized absolute difference]

Site name	Sample collection date	Plutonium-238		Plutonium-239+240		Americium-241	
		(pCi/L)	(NAD)	(pCi/L)	(NAD)	(pCi/L)	(NAD)
USGS 87	01-16-96	0.01±0.01 0.004±0.013	0.31	-0.007±0.012 0.003±0.013	0.57	-0.03±0.02 0.00±0.02	1.06
USGS 120	01-17-96	-0.007±0.012 -0.004±0.012	0.18	0.003±0.013 -0.011±0.012	0.79	0.014±0.019 0.00±0.02	0.51
USGS 34	04-02-96	0.004±0.016 0.016±0.022	0.44	0.02±0.02 0.03±0.02	0.35	-0.01±0.02 -0.02±0.02	0.35
USGS 117	04-16-96	0.012±0.021 0.003±0.015	0.35	0.012±0.018 -0.005±0.014	0.75	0.015±0.014 0.00±0.02	0.61
USGS 77	10-17-96	-0.004±0.013 -0.001±0.015	0.15	-0.008±0.013 -0.02±0.012	0.68	0.02±0.03 0.00±0.02	0.55
USGS 38	10-25-96	-0.003±0.012 -0.017±0.016	0.70	-0.001±0.012 -0.007±0.014	0.33	0.02±0.02 0.00±0.02	0.71
USGS 43	10-28-96	0.01±0.01 -0.007±0.012	0.92	0.006±0.012 0.003±0.013	0.17	-0.01±0.019 0.013±0.017	0.90
USGS 37	10-8-97	-0.001±0.017 -0.001±0.012	0.00	0.02±0.02 0.003±0.014	0.70	-0.01±0.02 -0.004±0.015	0.24
USGS 117	04-01-98	0.004±0.013 -0.003±0.012	0.40	0.001±0.013 0.005±0.013	0.22	0.01±0.02 -0.03±0.02	1.41
USGS 65	04-28-98	-0.014±0.014 -0.007±0.013	0.37	-0.001±0.013 -0.002±0.012	0.06	-0.005±0.017 -0.021±0.016	0.69
USGS 90	07-21-98	0.000±0.012 0.01±0.01	0.70	-0.009±0.012 0.00±0.01	0.32	-0.006±0.016 0.00±0.02	0.23
USGS 97	10-14-98	-0.012±0.012 -0.01±0.01	0.19	0.009±0.013 0.00±0.01	0.37	0.006±0.016 0.00±0.02	0.23
USGS 47	10-28-98	0.008±0.013 0.011±0.013	0.16	-0.001±0.012 0.006±0.012	0.41	-0.005±0.013 0.008±0.014	0.68
USGS 88	01-13-99	-0.004±0.009 -0.004±0.009	0.01	-0.008±0.008 0.00±0.01	0.31	0.00±0.01 -0.003±0.009	0.01
USGS 40	01-19-99	0.003±0.008 -0.004±0.008	0.06	-0.003±0.007 -0.0037±0.008	0.07	-0.01±0.01 0.02±0.01	2.47
CPP 1	04-26-99	-0.007±0.007 -0.004±0.008	0.28	-0.003±0.008 0.00±0.01	0.55	0.00±0.01 -0.02±0.01	1.27
USGS 34	10-14-99	0.000±0.006 0.01±0.02	0.48	0.003±0.006 0.013±0.016	0.59	0.004±0.009 -0.006±0.006	0.92
USGS 98	04-04-00	-0.009±0.005 0.004±0.012	1.00	-0.006±0.004 -0.009±0.006	0.42	0.009±0.008 0.016±0.016	0.39
USGS 119	04-06-00	-0.009±0.005 0.003±0.007	1.39	-0.003±0.005 -0.003±0.006	0.00	0.006±0.006 0.011±0.009	0.46
USGS 97	09-27-00	-0.004±0.007 0.006±0.011	0.77	-0.004±0.007 0.006±0.011	0.77	0.008±0.012 0.010±0.008	0.10
USGS 77	10-06-00	0.006±0.008	0.07	0.006±0.008	0.07	0.006±0.006	0.34

Table 4. Measured concentrations and normalized absolute differences for plutonium-238, plutonium-239+240, and americium-241 from replicate pairs collected from selected sites at the Idaho National Laboratory and vicinity, Idaho, 1996–2001.—Continued

[Locations of sites are shown in figures 1–4. Uncertainties are 1σ combined standard uncertainties. **Abbreviations:** pCi/L, picocuries per liter; NAD, normalized absolute difference]

Site name	Sample collection date	Plutonium-238		Plutonium-239+240		Americium-241	
		(pCi/L)	(NAD)	(pCi/L)	(NAD)	(pCi/L)	(NAD)
		0.007±0.013		0.007±0.013		0.01±0.01	
RWMC M1SA	01-16-01	0.004±0.007	0.78	0.004±0.008	0.09	0.006±0.009	0.24
		-0.0016±0.0016		0.0032±0.0032		0.0084±0.0045	
USGS 88	01-23-01	0.003±0.006	0.97	0.007±0.008	1.23	-0.003±0.009	0.42
		-0.004±0.004		-0.004±0.004		0.003±0.011	
USGS 84	04-30-01	0.003±0.005	0.00	0.006±0.006	0.33	-0.009±0.009	1.27
		0.003±0.006		0.003±0.007		0.009±0.011	
USGS 120	07-12-01	-0.003±0.005	1.41	0.006±0.004	0.47	0.003±0.009	0.00
		0.006±0.004		0.003±0.005		0.003±0.009	
RWMC M1SA	09-25-01	-0.00009±0.00011	1.03	0.003±0.003	0.79	-0.008±0.005	1.77
		0.003±0.003		-0.003±0.007		0.015±0.012	
USGS 38	10-11-01	0.011±0.011	0.32	-0.003±0.003	0.00	0.006±0.006	0.20
		0.006±0.011		-0.003±0.005		0.008±0.008	
USGS 43	10-22-01	0.006±0.013	0.18	-0.003±0.005	0.84	0.003±0.005	0.00
		0.011±0.0225		0.011±0.016		0.003±0.006	

Table 5. Measured concentrations and relative standard deviations for sodium, chloride, fluoride, and sulfate from replicate pairs collected from selected sites at the Idaho National Laboratory and vicinity, 1996–2001.

[Locations of sites are shown in <u>figures 1–4</u>. Concentrations in table are rounded results, but relative standard deviations (RSDs) were calculated using unrounded results. **Abbreviations:** mg/L, milligrams per liter; –, no data]

Site name	Sample collection date	Sodium		Chloride		Fluoride		Sulfate	
		(mg/L)	RSD (percent)	(mg/L)	RSD (percent)	(mg/L)	RSD (percent)	(mg/L)	RSD (percent)
USGS 103	01-02-96	–	–	15	0	–	–	–	–
		–		15		–		–	
USGS 125	01-09-96	13	6	14	0	–	–	–	–
		12		14		–		–	
RWMC Production	01-16-96	–	–	17	4	–	–	–	–
		–		18		–		–	
USGS 87	01-16-96	–	–	13	0	–	–	–	–
		–		13		–		–	
USGS 112	01-18-96	–	–	120	0	–	–	–	–
		–		120		–		–	
USGS 83	04-01-96	11	7	11	0	–	–	–	–
		10		11		–		–	
USGS 34	04-02-96	9.8	1	14	0	0.2	0	31	0
		9.9		14		0.2		31	
USGS 79	04-02-96	–	–	12	0	–	–	–	–
		–		12		–		–	
CFA 2	04-03-96	26	0	120	0	–	–	45	0
		26		120		–		45	
Atomic City	04-04-96	–	–	17	0	–	–	–	–
		–		17		–		–	
USGS 5	04-10-96	7.3	1	8.4	1	–	–	–	–
		7.2		8.5		–		–	
USGS 20	04-11-96	–	–	24	0	–	–	–	–
		–		24		–		–	
USGS 117	04-16-96	–	–	13	5	–	–	–	–
		–		14		–		–	
USGS 9	04-16-96	13	0	21	0	–	–	–	–
		13		21		–		–	
MTR Test	04-17-96	31	2	13	0	–	–	140	5
		30		13		–		150	
USGS 109	04-17-96	11	0	15	0	–	–	–	–
		11		15		–		–	

Table 5. Measured concentrations and relative standard deviations for sodium, chloride, fluoride, and sulfate from replicate pairs collected from selected sites at the Idaho National Laboratory and vicinity, 1996–2001.—Continued

[Locations of sites are shown in figures 1–4. Concentrations in table are rounded results, but relative standard deviations (RSDs) were calculated using unrounded results. **Abbreviations:** mg/L, milligrams per liter; –, no data]

Site name	Sample collection date	Sodium		Chloride		Fluoride		Sulfate	
		(mg/L)	RSD (percent)	(mg/L)	RSD (percent)	(mg/L)	RSD (percent)	(mg/L)	RSD (percent)
TRA Disposal	04-18-96	–	–	10	7	–	–	–	–
		–		11		–		–	
USGS 52	04-22-96	–	–	28	2	–	–	–	–
		–		29		–		–	
SITE 14	04-24-96	15	5	9.0	1	–	–	–	–
		14		8.9		–		–	
Leo Rogers 1	07-17-96	17	0	18	0	–	–	–	–
		17		18		–		–	
IET 1 Disposal	07-18-96	18	1	18	0	–	–	31	0
		18		18		–		31	
TRA 1	07-18-96	8.3	2	10	7	–	–	21	3
		8.5		11		–		20	
TRA 3	07-18-96	8.6	0	10	0	–	–	21	0
		8.6		10		–		21	
ANP 6	07-19-96	9.8	1	17	0	–	–	33	1
		9.9		17		–		33	
RWMC M3S	07-22-96	–	–	12	2	–	–	–	–
		–		13		–		–	
USGS 66	07-25-96	15	0	19	4	–	–	190	0
		15		20		–		190	
USGS 72	07-30-96	22	3	13	5	–	–	30	0
		21		14		–		30	
Big Lost River near Arco (BLR nr Arco)	10-07-96	–	–	5.8	5	–	–	–	–
		–		6.2		–		–	
Birch Creek at Blue Dome Inn near Reno (BC at Blue Dome nr Reno)	10-08-96	–	–	4.8	5	–	–	–	–
		–		4.5		–		–	
No Name 1	10-14-96	11	0	20	0	–	–	–	–
		11		20		–		–	
USGS 27	10-15-96	28	3	58	4	–	–	–	–
		27		61		–		–	
SPERT 1	10-16-96	13	0	21	3	–	–	–	–
		13		20		–		–	

Table 5. Measured concentrations and relative standard deviations for sodium, chloride, fluoride, and sulfate from replicate pairs collected from selected sites at the Idaho National Laboratory and vicinity, 1996–2001.—Continued

[Locations of sites are shown in figures 1–4. Concentrations in table are rounded results, but relative standard deviations (RSDs) were calculated using unrounded results. **Abbreviations:** mg/L, milligrams per liter; –, no data]

Site name	Sample collection date	Sodium		Chloride		Fluoride		Sulfate	
		(mg/L)	RSD (percent)	(mg/L)	RSD (percent)	(mg/L)	RSD (percent)	(mg/L)	RSD (percent)
USGS 77	10-17-96	33	2	140	0	0.2	0	32	0
		34		140		0.2		32	
USGS 48	10-21-96	20	3	41	3	–	–	28	2
		21		43		–		29	
PW 3	10-23-96	170	4	290	2	–	–	27	0
		180		300		–		27	
USGS 73	10-23-96	13	5	34	2	–	–	61	0
		14		33		–		61	
USGS 38	10-25-96	65	1	180	0	0.2	0	31	0
		66		180		0.2		31	
USGS 43	10-28-96	16	5	24	3	–	–	25	0
		15		25		–		25	
USGS 113	01-09-97	–	–	230	0	–	–	–	–
		–		230		–		–	
USGS 54	01-23-97	–	–	26	3	–	–	320	2
		–		27		–		330	
USGS 60	04-02-97	–	–	21	0	–	–	–	–
		–		21		–		–	
USGS 109	04-03-97	10	1	14	1	–	–	–	–
		11		14		–		–	
USGS 86	04-03-97	10	1	20	1	–	–	–	–
		11		20		–		–	
CPP 5	04-08-97	7.7	0	19	3	–	–	26	0
		7.7		19		–		26	
NPR TEST	04-08-97	7.6	1	13	1	–	–	–	–
		7.4		13		–		–	
USGS 41	04-08-97	–	–	21	0	–	–	–	–
		–		21		–		–	
USGS 42	04-08-97	–	–	26	0	–	–	–	–
		–		26		–		–	
Big Lost River at Experimental Dairy Farm near Howe (BLR at exp dairy farm nr Howe)	04-09-97	–	–	4.5	4	–	–	–	–
		–		4.3		–		–	

Table 5. Measured concentrations and relative standard deviations for sodium, chloride, fluoride, and sulfate from replicate pairs collected from selected sites at the Idaho National Laboratory and vicinity, 1996–2001.—Continued

[Locations of sites are shown in figures 1–4. Concentrations in table are rounded results, but relative standard deviations (RSDs) were calculated using unrounded results. **Abbreviations:** mg/L, milligrams per liter; –, no data]

Site name	Sample collection date	Sodium		Chloride		Fluoride		Sulfate	
		(mg/L)	RSD (percent)	(mg/L)	RSD (percent)	(mg/L)	RSD (percent)	(mg/L)	RSD (percent)
USGS 124	04-10-97	–	–	15	0	–	–	–	–
		–		15		–		–	
USGS 63	04-22-97	–	–	23	2	–	–	–	–
		–		22		–		–	
USGS 104	04-29-97	–	–	13	0	–	–	–	–
		–		13		–		–	
USGS 58	05-05-97	–	–	12	3	–	–	–	–
		–		12		–		–	
USGS 6	07-01-97	12	0	8.7	0	–	–	18	1
		13		8.6		–		18	
USGS 82	07-02-97	–	–	19	1	–	–	–	–
		–		19		–		–	
USGS 32	07-08-97	20	4	60	3	–	–	44	5
		19		57		–		41	
CWP 1	07-15-97	–	–	26	0	–	–	320	0
		–		26		–		320	
Badging Facility	07-22-97	9.5	1	17	3	–	–	20	1
		9.6		16		–		20	
USGS 108	10-07-97	11	1	14	0	–	–	–	–
		11		14		–		–	
USGS 39	10-07-97	7.5	3	10	0	–	–	31	1
		7.8		10		–		31	
USGS 37	10-08-97	48	1	140	0	–	–	30	0
		49		140		–		30	
USGS 8	100-8-97	6.7	0	7.5	0	–	–	–	–
		6.7		7.5		–		–	
USGS 101	10-16-97	14	0	8.5	0	–	–	–	–
		14		8.5		–		–	
PW 8	10-21-97	19	2	24	1	–	–	283	1
		18		24		–		279	
SITE 4	10-21-97	7.9	0	9.1	1	–	–	19	4
		7.9		9.3		–		18	
USGS 67	10-21-97	52	2	160	0	–	–	30	1
		54		160		–		30	

Table 5. Measured concentrations and relative standard deviations for sodium, chloride, fluoride, and sulfate from replicate pairs collected from selected sites at the Idaho National Laboratory and vicinity, 1996–2001.—Continued

[Locations of sites are shown in figures 1–4. Concentrations in table are rounded results, but relative standard deviations (RSDs) were calculated using unrounded results. **Abbreviations:** mg/L, milligrams per liter; –, no data]

Site name	Sample collection date	Sodium		Chloride		Fluoride		Sulfate	
		(mg/L)	RSD (percent)	(mg/L)	RSD (percent)	(mg/L)	RSD (percent)	(mg/L)	RSD (percent)
USGS 115	10-22-97	15	0	40	0	–	–	22	0
		15		40		–		22	
ANP 9	10-27-97	14	0	12	1	–	–	–	–
		14		12		–		–	
USGS 51	10-27-97	27	1	140	0	–	–	28	0
		27		140		–		28	
USGS 119	01-13-98	–	–	9.1	4	–	–	–	–
		–		8.6		–		–	
USGS 104	01-15-98	–	–	13	3	–	–	–	–
		–		12		–		–	
USGS 114	01-22-98	–	–	71	0	–	–	–	–
		–		72		–		–	
PW 1	01-26-98	–	–	225	1	–	–	–	–
		–		222		–		–	
USGS 5	03-31-98	7.2	0	9.0	0	–	–	–	–
		7.3		9.0		–		–	
USGS 117	04-01-98	–	–	13	1	–	–	–	–
		–		13		–		–	
USGS 121	04-02-98	–	–	11	1	–	–	–	–
		–		11		–		–	
USGS 106	04-06-98	–	–	15	3	–	–	–	–
		–		14		–		–	
USGS 111	04-06-98	–	–	14	117	–	–	–	–
		–		153		–		–	
USGS 14	04-07-98	–	–	21	1	–	–	–	–
		–		20		–		–	
USGS 19	04-08-98	11	1	9.0	1	–	–	–	–
		11		9.1		–		–	
USGS 7	04-13-98	22	6	8.9	1	–	–	–	–
		24		8.8		–		–	
Highway 3	04-20-98	5.8	0	5.8	4	–	–	–	–
		5.8		6.2		–		–	
Arbor Test	04-22-98	–	–	144	117	–	–	–	–
		–		14		–		–	

Table 5. Measured concentrations and relative standard deviations for sodium, chloride, fluoride, and sulfate from replicate pairs collected from selected sites at the Idaho National Laboratory and vicinity, 1996–2001.—Continued

[Locations of sites are shown in figures 1–4. Concentrations in table are rounded results, but relative standard deviations (RSDs) were calculated using unrounded results. **Abbreviations:** mg/L, milligrams per liter; –, no data]

Site name	Sample collection date	Sodium		Chloride		Fluoride		Sulfate	
		(mg/L)	RSD (percent)	(mg/L)	RSD (percent)	(mg/L)	RSD (percent)	(mg/L)	RSD (percent)
USGS 65	04-28-98	14	1	18	1	–	–	143	0
		14		18		–		144	
USGS 85	04-28-98	–	–	46	0	–	–	–	–
		–		46		–		–	
USGS 69	07-01-98	12	6	20	2	–	–	154	1
		11		19		–		152	
RWMC M3S	07-07-98	–	–	12	0	–	–	–	–
		–		12		–		–	
USGS 18	07-20-98	12	0	10	2	–	–	24	0
		12		10		–		24	
TRA 4	07-21-98	8.3	10	10	3	–	–	19	0
		7.2		10		–		19	
USGS 90	07-21-98	–	–	17	1	–	–	–	–
		–		17		–		–	
USGS 2	07-22-98	17	1	16	0	–	–	13	0
		17		16		–		13	
USGS 36	07-22-98	–	–	68	0	–	–	–	–
		–		68		–		–	
CPP 2	09-30-98	7.4	0	14	0	–	–	–	–
		7.4		14		–		–	
Mud Lake near Terreton (ML nr Terreton)	10-06-98	–	–	8.0	2	–	–	–	–
		–		8.2		–		–	
USGS 97	10-14-98	15	1	37	1	–	–	34	0
		14		37		–		34	
USGS 35	10-15-98	7.8	1	9.4	2	–	–	26	0
		7.7		9.2		–		26	
USGS 1	10-20-98	15	1	13	5	–	–	–	–
		15		14		–		–	
USGS 110A	10-20-98	17	0	19	2	–	–	–	–
		17		20		–		–	
USGS 123	10-20-98	48	1	111	1	–	–	30	0
		48		113		–		30	
WS INEL 1	10-20-98	12	1	44	3	–	–	32	0
		12		46		–		32	

Table 5. Measured concentrations and relative standard deviations for sodium, chloride, fluoride, and sulfate from replicate pairs collected from selected sites at the Idaho National Laboratory and vicinity, 1996–2001.—Continued

[Locations of sites are shown in figures 1–4. Concentrations in table are rounded results, but relative standard deviations (RSDs) were calculated using unrounded results. **Abbreviations:** mg/L, milligrams per liter; –, no data]

Site name	Sample collection date	Sodium		Chloride		Fluoride		Sulfate	
		(mg/L)	RSD (percent)	(mg/L)	RSD (percent)	(mg/L)	RSD (percent)	(mg/L)	RSD (percent)
USGS 76	10-22-98	9.5	1	12	1	–	–	26	0
		9.4		12		–		26	
PW 6	10-27-98	112	0	160	3	–	–	14	0
		112		167		–		14	
USGS 47	10-28-98	15	4	26	0	–	–	25	1
		14		26		–		25	
USGS 68	12-01-98	662	1	27	2	–	–	2,280	2
		674		26		–		2,340	
USGS 88	01-13-99	–	–	84	0	–	–	–	–
		–		84		–		–	
USGS 40	01-19-99	–	–	17	1	–	–	–	–
		–		17		–		–	
USGS 99	03-31-99	–	–	20	0	–	–	–	–
		–		20		–		–	
USGS 79	04-01-99	–	–	13	1	–	–	–	–
		–		13		–		–	
USGS 83	04-01-99	9.5	6	11	2	–	–	–	–
		10		10		–		–	
USGS 12	04-05-99	13	0	16	1	–	–	25	0
		13		16		–		25	
EBR 1	04-13-99	8.0	1	6.8	1	–	–	–	–
		8.0		6.9		–		–	
USGS 61	04-13-99	–	–	20	1	–	–	–	–
		–		20		–		–	
P AND W 2	04-14-99	6.8	7	7.1	0	–	–	–	–
		6.2		7.1		–		–	
USGS 100	04-19-99	–	–	16	3	–	–	–	–
		–		17		–		–	
Big Lost River below INL Diversion near Arco (BLR blw INL div nr Arco)	04-20-99	–	–	5.0	0	–	–	–	–
		–		4.9		–		–	
PW 2	04-22-99	–	–	200	0	–	–	–	–
		–		200		–		–	
USGS 46	04-22-99	–	–	17	1	–	–	–	–
		–		17		–		–	

Table 5. Measured concentrations and relative standard deviations for sodium, chloride, fluoride, and sulfate from replicate pairs collected from selected sites at the Idaho National Laboratory and vicinity, 1996–2001.—Continued

[Locations of sites are shown in figures 1–4. Concentrations in table are rounded results, but relative standard deviations (RSDs) were calculated using unrounded results. **Abbreviations:** mg/L, milligrams per liter; –, no data]

Site name	Sample collection date	Sodium		Chloride		Fluoride		Sulfate	
		(mg/L)	RSD (percent)	(mg/L)	RSD (percent)	(mg/L)	RSD (percent)	(mg/L)	RSD (percent)
CPP 1	04-26-99	7.8	1	15	3	0.2	3	23	0
		7.7		14		0.2		23	
CWP 3	06-30-99	–	–	24	0	–	–	295	0
		–		24		–		293	
USGS 22	07-07-99	22	3	63	67	–	–	–	–
		23		23		–		–	
SPERT 1	07-08-99	15	1	24	0	–	–	–	–
		15		24		–		–	
RWMC M7S	07-12-99	–	–	12	0	–	–	–	–
		–		12		–		–	
USGS 15	07-13-99	23	1	35	1	–	–	35	2
		23		35		–		34	
TRA 1	07-21-99	8.5	0	11	1	–	–	18	0
		8.6		11		–		18	
PW 4	10-06-99	152	0	316	2	–	–	29	0
		152		308		–		29	
USGS 114	10-06-99	19	2	83	2	–	–	27	0
		20		81		–		27	
USGS 11	10-07-99	9.4	13	11	4	–	–	–	–
		7.8		11		–		–	
Big Lost River below Mackay Reservoir (BLR blw Mackay Reservoir)	10-07-99	–	–	3.7	19	–	–	–	–
		–		2.8		–		–	
USGS 20	10-14-99	8.2	0	30	1	–	–	20	1
		8.3		31		–		21	
USGS 34	10-14-99	9.6	0	15	0	0.2	0	24	1
		9.6		15		0.2		24	
CFA 1	10-20-99	22	4	85	0	–	–	29	0
		20		86		–		29	
USGS 50	10-20-99	63	4	57	1	–	–	39	0
		66		58		–		39	
USGS 58	10-21-99	9.0	4	11	0	–	–	27	0
		9.6		11		–		27	

Table 5. Measured concentrations and relative standard deviations for sodium, chloride, fluoride, and sulfate from replicate pairs collected from selected sites at the Idaho National Laboratory and vicinity, 1996–2001.—Continued

[Locations of sites are shown in figures 1–4. Concentrations in table are rounded results, but relative standard deviations (RSDs) were calculated using unrounded results. **Abbreviations:** mg/L, milligrams per liter; –, no data]

Site name	Sample collection date	Sodium		Chloride		Fluoride		Sulfate	
		(mg/L)	RSD (percent)	(mg/L)	RSD (percent)	(mg/L)	RSD (percent)	(mg/L)	RSD (percent)
PW 5	10-25-99	163	0	348	1	–	–	27	0
		164		344		–		27	
USGS 105	10-25-99	14	1	13	5	–	–	–	–
		14		12		–		–	
USGS 121	10-25-99	6.0	12	11	1	–	–	22	1
		7.2		11		–		22	
USGS 52	10-25-99	12	2	16	32	–	–	24	11
		12		25		–		27	
USGS 23	10-27-99	8.9	5	10	1	–	–	–	–
		9.5		10		–		–	
RWMC Production	01-13-00	–	–	18	0	–	–	–	–
		–		18		–		–	
CFA 2	01-19-00	29	1	125	2	–	–	47	0
		28		122		–		47	
USGS 89	01-20-00	–	–	37	0	–	–	–	–
		–		38		–		–	
PSTF Test	04-04-00	6.8	3	6.1	10	–	–	–	–
		7.0		7.0		–		–	
USGS 98	04-04-00	10	2	14	2	–	–	21	0
		10		14		–		21	
USGS 12	04-05-00	12	0	13	2	–	–	22	0
		12		13		–		22	
USGS 119	04-06-00	–	–	8.7	1	–	–	–	–
		–		8.8		–		–	
MTR Test	04-10-00	10	1	9.2	1	–	–	25	1
		9.9		9.2		–		24	
Atomic City	04-11-00	–	–	16	2	–	–	–	–
		–		16		–		–	
Little Lost River Nr Howe (LLR nr Howe)	04-11-00	–	–	9.3	8	–	–	–	–
		–		10		–		–	
USGS 125	04-11-00	13	4	12	1	–	–	–	–
		12		12		–		–	
USGS 26	04-12-00	15	1	13	0	–	–	–	–
		16		13		–		–	

Table 5. Measured concentrations and relative standard deviations for sodium, chloride, fluoride, and sulfate from replicate pairs collected from selected sites at the Idaho National Laboratory and vicinity, 1996–2001.—Continued

[Locations of sites are shown in figures 1–4. Concentrations in table are rounded results, but relative standard deviations (RSDs) were calculated using unrounded results. **Abbreviations:** mg/L, milligrams per liter; –, no data]

Site name	Sample collection date	Sodium		Chloride		Fluoride		Sulfate	
		(mg/L)	RSD (percent)	(mg/L)	RSD (percent)	(mg/L)	RSD (percent)	(mg/L)	RSD (percent)
USGS 27	04-12-00	28	6	58	3	–	–	–	–
		30		56		–		–	
USGS 102	07-10-00	15	0	31	1	–	–	31	0
		15		30		–		31	
USGS 31	07-10-00	15	1	24	0	–	–	29	1
		15		23		–		29	
Badging Facility	07-13-00	9.9	0	16	0	–	–	22	0
		9.9		16		–		22	
RWMC Production	07-13-00	–	–	17	1	–	–	–	–
		–		17		–		–	
SITE 19	07-13-00	9.4	1	13	0	–	–	35	1
		9.5		13		–		36	
CWP 2	07-17-00	–	–	17	0	–	–	123	0
		–		17		–		123	
AREA 2	07-18-00	16	1	16	1	–	–	17	0
		16		16		–		17	
USGS 66	07-25-00	15	3	18	1	–	–	197	3
		15		18		–		190	
USGS 62	09-27-00	18	1	22	1	–	–	262	1
		18		22		–		260	
USGS 97	09-27-00	16	0	31	1	–	–	32	0
		16		32		–		32	
USGS 82	09-28-00	10	0	16	5	–	–	21	0
		10		17		–		21	
USGS 58	10-04-00	10	0	11	2	–	–	27	0
		10		11		–		27	
USGS 60	10-04-00	19	2	23	1	–	–	282	0
		18		24		–		282	
USGS 11	10-05-00	8.4	0	10	1	–	–	–	–
		8.4		10		–		–	
USGS 124	10-05-00	9.8	2	15	1	–	–	–	–
		9.6		15		–		–	
USGS 57	10-05-00	44	1	91	1	–	–	29	1
		43		92		–		29	

Table 5. Measured concentrations and relative standard deviations for sodium, chloride, fluoride, and sulfate from replicate pairs collected from selected sites at the Idaho National Laboratory and vicinity, 1996–2001.—Continued

[Locations of sites are shown in figures 1–4. Concentrations in table are rounded results, but relative standard deviations (RSDs) were calculated using unrounded results. **Abbreviations:** mg/L, milligrams per liter; –, no data]

Site name	Sample collection date	Sodium		Chloride		Fluoride		Sulfate	
		(mg/L)	RSD (percent)	(mg/L)	RSD (percent)	(mg/L)	RSD (percent)	(mg/L)	RSD (percent)
USGS 77	10-06-00	39	1	149	1	0.2	4	32	0
		38		147		0.2		32	
USGS 50	10-10-00	60	0	57	0	–	–	41	0
		61		57		–		41	
Birch Creek at Blue Dome Inn near Reno (BC at Blue Dome nr Reno)	10-17-00	–	–	5.4	0	–	–	–	–
		–		5.4		–		–	
RWMC M13S	10-17-00	11	0	5.6	0	–	–	–	–
		11		5.6		–		–	
USGS 104	10-23-00	9.2	4	12	3	–	–	–	–
		8.8		12		–		–	
Atomic City	10-24-00	18	5	17	0	–	–	–	–
		17		17		–		–	
USGS 116	01-10-01	–	–	101	2	–	–	–	–
		–		98		–		–	
USGS 88	01-23-01	–	–	79	1	–	–	–	–
		–		80		–		–	
USGS 73	04-02-01	–	–	91	0	–	–	–	–
		–		92		–		–	
USGS 103	04-04-01	13	4	15	0	–	–	–	–
		13		15		–		–	
USGS 17	04-04-01	7.7	5	5.3	0	–	–	–	–
		7.2		5.4		–		–	
SITE 14	04-05-01	14	1	8.3	0	–	–	–	–
		15		8.3		–		–	
Birch Creek at Blue Dome Inn near Reno (BC at Blue Dome nr Reno)	04-11-01	–	–	5.1	2	–	–	–	–
		–		5.0		–		–	
USGS 9	04-11-01	11	4	17	0	–	–	–	–
		12		17		–		–	
PW 4	04-12-01	–	–	152	0	–	–	–	–
		–		153		–		–	
RWMC M14S	04-17-01	8.9	2	15	2	–	–	–	–
		8.6		16		–		–	

Table 5. Measured concentrations and relative standard deviations for sodium, chloride, fluoride, and sulfate from replicate pairs collected from selected sites at the Idaho National Laboratory and vicinity, 1996–2001.—Continued

[Locations of sites are shown in figures 1–4. Concentrations in table are rounded results, but relative standard deviations (RSDs) were calculated using unrounded results. **Abbreviations:** mg/L, milligrams per liter; –, no data]

Site name	Sample collection date	Sodium		Chloride		Fluoride		Sulfate	
		(mg/L)	RSD (percent)	(mg/L)	RSD (percent)	(mg/L)	RSD (percent)	(mg/L)	RSD (percent)
USGS 127	04-18-01	8.1 7.6	4	15 15	0	– –	–	– –	–
USGS 41	04-23-01	– –	–	18 18	1	– –	–	– –	–
USGS 44	04-24-01	– –	–	10 10	1	– –	–	– –	–
CFA LF 2-10	04-30-01	12 12	1	23 24	2	– –	–	– –	–
USGS 84	04-30-01	8.2 7.9	2	8.2 8.3	1	– –	–	33 33	0
USGS 45	05-01-01	– –	–	16 16	0	– –	–	– –	–
Site 9	07-05-01	12 13	1	14 14	0	– –	–	24 25	1
USGS 126A	07-10-01	8.1 8.4	2	8.2 8.0	1	– –	–	– –	–
USGS 126B	07-10-01	8.0 7.8	2	8.1 8.3	2	– –	–	– –	–
USGS 120	07-12-01	– –	–	19 19	1	– –	–	– –	–
USGS 29	07-16-01	21 21	0	28 28	1	– –	–	17 17	0
TRA 3	07-19-01	9.6 9.0	4	9.5 9.8	2	– –	–	22 22	0
Site 17	07-30-01	11 11	0	13 13	0	– –	–	– –	–
Birch Creek at Blue Dome Inn near Reno (BC at Blue Dome nr Reno)	10-04-01	– –	–	5.8 5.7	2	– –	–	– –	–
PW 8	10-09-01	15 14	1	20 20	0	– –	–	179 179	0
USGS 38	10-11-01	45 45	0	100 100	0	0.21 0.26	[1]16	28 29	2

Table 5. Measured concentrations and relative standard deviations for sodium, chloride, fluoride, and sulfate from replicate pairs collected from selected sites at the Idaho National Laboratory and vicinity, 1996–2001.—Continued

[Locations of sites are shown in figures 1–4. Concentrations in table are rounded results, but relative standard deviations (RSDs) were calculated using unrounded results. **Abbreviations:** mg/L, milligrams per liter; –, no data]

Site name	Sample collection date	Sodium		Chloride		Fluoride		Sulfate	
		(mg/L)	RSD (percent)	(mg/L)	RSD (percent)	(mg/L)	RSD (percent)	(mg/L)	RSD (percent)
Arbor Test	10-15-01	–	–	15	0	–	–	–	–
		–		15		–		–	
USGS 106	10-17-01	7.5	1	15	2	–	–	–	–
		7.5		15		–		–	
USGS 42	10-17-01	8.6	0	18	5	–	–	24	1
		8.7		17		–		24	
USGS 43	10-22-01	14	1	23	0	–	–	23	0
		14		23		–		23	

[1]Replicate pair concentrations were within ±0.5 times the reporting level

Table 6. Measured concentrations, relative standard deviations, and acceptable or not acceptable reproducibility for ammonia, nitrate + nitrite, nitrite, and orthophosphate from replicate pairs collected from selected sites at the Idaho National Laboratory and vicinity, Idaho, 1996–2001.

[Locations of sites are shown in figures 1–4. Concentrations in table are rounded results, but relative standard deviations (RSDs) were calculated using unrounded results. **Reproducibility:** A, acceptable; N, not acceptable. **Abbreviations:** mg/L, milligrams per liter; N, nitrogen; P, phosphorous; E, estimated; <, less than; NC, not calculated because one or both concentrations were below the reporting level]

Site name	Sample collection date	Ammonia			Nitrate + nitrite			Nitrite			Orthophosphate		
		(mg/L as N)	RSD (percent)	Reproduc-ibility	(mg/L as N)	RSD (percent)	Reproduc-ibility	(mg/L as N)	RSD (percent)	Reproduc-ibility	(mg/L as P)	RSD (percent)	Reproduc-ibility
USGS 125	01-09-96	<0.015 <0.015	NC	A	0.55 0.56	1	A	<0.01 <0.01	NC	A	0.01 0.01	0	A
USGS 83	04-01-96	<0.015 <0.015	NC	A	0.62 0.63	1	A	<0.01 <0.01	NC	A	<0.01 <0.01	NC	A
USGS 34	04-02-96	<0.015 <0.015	NC	A	1.1 1.1	0	A	<0.01 <0.01	NC	A	0.02 0.02	0	A
USGS 5	04-10-96	<0.015 <0.015	NC	A	0.43 0.43	0	A	<0.01 <0.01	NC	A	<0.01 <0.01	NC	A
USGS 9	04-16-96	<0.015 <0.015	NC	A	0.57 0.6	4	A	<0.01 <0.01	NC	A	0.02 0.02	0	A
USGS 109	04-17-96	<0.015 <0.015	NC	A	0.52 0.54	3	A	<0.01 <0.01	NC	A	<0.01 <0.01	NC	A
SITE 14	04-24-96	0.02 <0.015	NC	A	0.54 0.54	0	A	<0.01 <0.01	NC	A	<0.01 <0.01	NC	A
IET 1 Disposal	07-18-96	0.74 0.75	1	A	0.73 0.73	0	A	0.02 0.02	0	A	0.21 0.22	3	A
ANP 6	07-19-96	0.03 0.03	0	A	0.91 0.91	0	A	<0.01 <0.01	NC	A	0.03 0.03	0	A
RWMC M3S	07-22-96	0.03 0.06	47	N	0.8 0.8	0	A	<0.01 <0.01	NC	A	0.02 0.03	28	N
No Name 1	10-14-96	0.02 0.02	0	A	0.56 0.58	2	A	<0.01 <0.01	NC	A	0.02 0.02	0	A
USGS 27	10-15-96	<0.015 <0.015	NC	A	2.6 2.6	0	A	0.02 <0.01	NC	N	<0.01 <0.01	NC	A
SPERT 1	10-16-96	<0.015 <0.015	NC	A	1 1.1	7	A	0.01 0.02	47	N	0.02 0.02	0	A
USGS 77	10-17-96	<0.015 <0.015	NC	A	4.3 4.3	0	A	0.01 0.01	0	A	0.02 0.02	0	A
USGS 48	10-21-96	<0.015 <0.015	NC	A	5.1 5.2	1	A	<0.01 0.02	NC	N	0.03 0.03	0	A
USGS 38	10-25-96	0.04 0.04	0	A	2.9 3	2	A	0.02 0.02	0	A	0.02 0.02	0	A
USGS 43	10-28-96	0.02 0.02	0	A	5.1 5.1	0	A	0.02 0.02	0	A	0.01 0.03	71	N
USGS 109	04-03-97	0.02 0.02	4	A	0.63 0.61	2	A	<0.01 <0.01	NC	A	<0.01 <0.01	NC	A
USGS 86	04-03-97	0.02 0.02	11	A	1.5 1.5	1	A	<0.01 <0.01	NC	A	0.02 0.02	5	A
NPR Test	04-08-97	<0.015 <0.015	NC	A	0.62 0.64	2	A	<0.01 <0.01	NC	A	0.02 0.02	3	A
USGS 6	07-01-97	0.03 <0.015	NC	N	0.52 0.52	0	A	<0.01 <0.01	NC	A	0.02 0.02	13	A

Table 6. Measured concentrations, relative standard deviations, and acceptable or not acceptable reproducibility for ammonia, nitrate + nitrite, nitrite, and orthophosphate from replicate pairs collected from selected sites at the Idaho National Laboratory and vicinity, Idaho, 1996–2001.—Continued

[Locations of sites are shown in figures 1–4. Concentrations in table are rounded results, but relative standard deviations (RSDs) were calculated using unrounded results. **Reproducibility:** A, acceptable; N, not acceptable. **Abbreviations:** mg/L, milligrams per liter; N, nitrogen; P, phosphorous; E, estimated; <, less than; NC, not calculated because one or both concentrations were below the reporting level]

Site name	Sample collection date	Ammonia			Nitrate + nitrite			Nitrite			Orthophosphate		
		(mg/L as N)	RSD (percent)	Reproduc-ibility	(mg/L as N)	RSD (percent)	Reproduc-ibility	(mg/L as N)	RSD (percent)	Reproduc-ibility	(mg/L as P)	RSD (percent)	Reproduc-ibility
USGS 32	07-08-97	<0.015 0.024	NC	N	1.72 1.79	3	A	<0.01 <0.01	NC	A	0.01 0.03	54	N
Badging Facility	07-22-97	<0.015 <0.015	NC	A	0.73 0.73	0	A	<0.01 <0.01	NC	A	0.02 0.02	8	A
USGS 107	10-01-97	<0.015 <0.015	NC	A	1.1 1.1	0	A	<0.01 <0.01	NC	A	0.01 0.01	0	A
USGS 108	10-07-97	0.02 <0.015	NC	A	0.74 0.74	1	A	<0.01 <0.01	NC	A	0.03 0.01	50	N
USGS 39	10-07-97	<0.015 <0.015	NC	A	0.84 0.85	1	A	<0.01 <0.01	NC	A	0.02 0.02	0	A
USGS 37	10-08-97	<0.015 <0.015	NC	A	2.9 2.9	1	A	<0.01 <0.01	NC	A	0.02 0.02	9	A
USGS 8	10-08-97	<0.015 <0.015	NC	A	0.92 0.93	1	A	<0.01 <0.01	NC	A	0.01 0.02	5	A
USGS 101	10-16-97	0.13 <0.015	NC	N	0.87 0.85	1	A	<0.01 <0.01	NC	A	0.01 <0.01	NC	A
USGS 67	10-21-97	<0.015 <0.015	NC	A	3.3 3.3	1	A	<0.01 <0.01	NC	A	0.01 0.01	5	A
USGS 115	10-22-97	0.04 <0.015	NC	N	1.4 1.4	1	A	<0.01 <0.01	NC	A	<0.01 0.01	NC	A
ANP 9	10-27-97	<0.015 <0.015	NC	A	0.92 0.92	0	A	<0.01 <0.01	NC	A	<0.01 <0.01	NC	A
USGS 51	10-27-97	<0.015 <0.015	NC	A	3.5 3.5	1	A	<0.01 <0.01	NC	A	<0.01 <0.01	NC	A
USGS 5	03-31-98	0.03 0.04	4	A	0.46 0.48	3	A	<0.01 <0.01	NC	A	<0.01 0.01	NC	A
USGS 19	04-08-98	0.02 0.02	10	A	0.87 0.87	1	A	<0.01 <0.01	NC	A	<0.01 0.01	NC	A
USGS 7	04-13-98	0.02 <0.02	NC	A	0.40 0.39	1	A	<0.01 <0.01	NC	A	0.02 0.01	14	A
Highway 3	04-20-98	0.03 0.03	3	A	0.37 0.37	0	A	<0.01 <0.01	NC	A	0.02 0.02	3	A
USGS 65	04-28-98	0.05 0.08	26	N	1.5 1.5	1	A	<0.01 <0.01	NC	A	0.02 0.02	17	A[1]
RWMC M3S	07-07-98	0.03 0.03	2	A	0.80 0.78	2	A	<0.01 <0.01	NC	A	0.03 0.03	0	A
USGS 18	07-20-98	0.03 <0.02	NC	A	0.57 0.62	6	A	<0.01 <0.01	NC	A	0.02 0.02	7	A
USGS 2	07-22-98	0.05 0.05	5	A	1.3 1.4	1	A	<0.01 <0.01	NC	A	0.02 0.02	7	A
CPP 2	09-30-98	<0.02 <0.02	NC	A	0.99 1.0	1	A	<0.01 <0.01	NC	A	0.02 0.02	10	A

Table 6.   Measured concentrations, relative standard deviations, and acceptable or not acceptable reproducibility for ammonia, nitrate + nitrite, nitrite, and orthophosphate from replicate pairs collected from selected sites at the Idaho National Laboratory and vicinity, Idaho, 1996–2001.—Continued

[Locations of sites are shown in figures 1–4. Concentrations in table are rounded results, but relative standard deviations (RSDs) were calculated using unrounded results. **Reproducibility:** A, acceptable; N, not acceptable. **Abbreviations:** mg/L, milligrams per liter; N, nitrogen; P, phosphorous; E, estimated; <, less than; NC, not calculated because one or both concentrations were below the reporting level]

Site name	Sample collection date	Ammonia			Nitrate + nitrite			Nitrite			Orthophosphate		
		(mg/L as N)	RSD (percent)	Reproduc- ibility	(mg/L as N)	RSD (percent)	Reproduc- ibility	(mg/L as N)	RSD (percent)	Reproduc- ibility	(mg/L as P)	RSD (percent)	Reproduc- ibility
USGS 97	10-14-98	<0.02 <0.02	NC	A	2.4 2.5	0	A	<0.01 <0.01	NC	A	0.03 0.03	5	A
USGS 35	10-15-98	0.02 <0.02	NC	A	0.76 0.77	1	A	<0.01 <0.01	NC	A	0.02 0.02	3	A
USGS 1	10-20-98	0.02 <0.02	NC	A	0.93 4.9	97	N	<0.01 <0.01	NC	A	<0.01 0.02	NC	N
USGS 110A	10-20-98	<0.02 ⁺0.02	NC	A	1.2 1.3	0	A	<0.01 <0.01	NC	A	<0.01 <0.01	NC	A
USGS 123	10-20-98	<0.02 <0.02	NC	A	0.94 4.9	96	N	<0.01 <0.01	NC	A	<0.01 0.02	NC	N
USGS 76	10-22-98	0.02 0.02	4	A	1.1 1.1	1	A	0.01 <0.01	NC	A	0.03 0.03	3	A
USGS 47	10-28-98	0.03 <0.02	NC	N	6.0 5.8	2	A	<0.01 <0.01	NC	A	0.02 0.03	21	N
USGS 83	04-01-99	<0.02 <0.02	NC	A	0.49 0.50	1	A	<0.01 <0.01	NC	A	<0.01 0.01	NC	A
USGS 12	04-05-99	<0.02 <0.02	NC	A	1.0 1.0	0	A	<0.01 <0.01	NC	A	0.02 0.02	6	A
EBR 1	04-13-99	<0.02 <0.02	NC	A	0.37 0.38	2	A	<0.01 <0.01	NC	A	0.01 0.02	10	A
P AND W 2	04-14-99	<0.02 <0.02	NC	A	0.34 0.34	1	A	<0.01 <0.01	NC	A	0.02 0.02	18	A[1]
SPERT 1	07-08-99	0.03 0.03	2	A	1.2 1.2	0	A	<0.01 <0.01	NC	A	0.03 0.02	11	A
RWMC M7S	07-12-99	<0.02 <0.02	NC	A	0.69 0.70	1	A	<0.01 <0.01	NC	A	0.02 0.02	8	A
USGS 15	07-13-99	<0.02 <0.02	NC	A	2.3 2.3	0	A	<0.01 <0.01	NC	A	0.02 0.02	0	A
USGS 114	10-06-99	<0.02 <0.02	NC	A	3.6 3.6	0	A	<0.01 <0.01	NC	A	0.01 0.01	13	A
USGS 11	10-07-99	<0.02 <0.02	NC	A	0.62 0.62	1	A	<0.01 <0.01	NC	A	0.01 0.01	7	A
USGS 20	10-14-99	<0.02 <0.02	NC	A	1.4 1.4	0	A	<0.01 <0.01	NC	A	0.01 0.02	20	A[1]
USGS 34	10-14-99	<0.02 <0.02	NC	A	0.98 0.99	1	A	<0.01 <0.01	NC	A	0.02 0.02	0	A
CFA 1	10-20-99	<0.02 <0.02	NC	A	3.3 3.4	1	A	<0.01 <0.01	NC	A	0.01 0.01	0	A
USGS 50	10-20-99	<0.02 <0.02	NC	A	58 58	0	A	<0.01 <0.01	NC	A	0.04 0.04	0	A
USGS 105	10-25-99	<0.02 <0.02	NC	A	0.62 0.62	0	A	<0.01 <0.01	NC	A	<0.01 <0.01	NC	A

Table 6. Measured concentrations, relative standard deviations, and acceptable or not acceptable reproducibility for ammonia, nitrate + nitrite, nitrite, and orthophosphate from replicate pairs collected from selected sites at the Idaho National Laboratory and vicinity, Idaho, 1996–2001.—Continued

[Locations of sites are shown in figures 1–4. Concentrations in table are rounded results, but relative standard deviations (RSDs) were calculated using unrounded results. **Reproducibility:** A, acceptable; N, not acceptable. **Abbreviations:** mg/L, milligrams per liter; N, nitrogen; P, phosphorous; E, estimated; <, less than; NC, not calculated because one or both concentrations were below the reporting level]

Site name	Sample collection date	Ammonia			Nitrate + nitrite			Nitrite			Orthophosphate		
		(mg/L as N)	RSD (percent)	Reproduc-ibility	(mg/L as N)	RSD (percent)	Reproduc-ibility	(mg/L as N)	RSD (percent)	Reproduc-ibility	(mg/L as P)	RSD (percent)	Reproduc-ibility
USGS 121	10-25-99	<0.02	NC	A	0.74	0	A	<0.01	NC	A	<0.01	NC	A
		<0.02			0.75			<0.01			0.01		
USGS 52	10-25-99	<0.02	NC	A	2.2	0	A	<0.01	NC	A	0.02	20	A[1]
		<0.02			2.2			<0.01			0.02		
USGS 23	10-27-99	<0.02	NC	A	0.82	0	A	<0.01	NC	A	<0.01	NC	A
		<0.02			0.82			<0.01			<0.01		
PSTF Test	04-04-00	<0.02	NC	A	0.59	0	A	<0.01	NC	A	0.02	7	A
		<0.02			0.59			<0.01			0.02		
USGS 98	04-04-00	<0.02	NC	A	1.2	0	A	<0.01	NC	A	0.01	25	N
		<0.02			1.2			<0.01			0.02		
USGS 12	04-05-00	<0.02	NC	A	0.90	0	A	<0.01	NC	A	0.02	6	A
		<0.02			0.90			<0.01			0.02		
USGS 125	04-11-00	<0.02	NC	A	0.60	0	A	<0.01	NC	A	<0.01	NC	A
		<0.02			0.60			<0.01			<0.01		
USGS 26	04-12-00	<0.02	NC	A	0.86	1	A	<0.01	NC	A	<0.01	NC	A
		<0.02			0.88			<0.01			<0.01		
USGS 27	04-12-00	<0.02	NC	A	2.6	0	A	<0.01	NC	A	<0.01	NC	A
		<0.02			2.6			<0.01			<0.01		
USGS 102	07-10-00	<0.02	NC	A	1.9	1	A	<0.01	NC	A	0.02	4	A
		<0.02			1.9			<0.01			0.02		
USGS 31	07-10-00	<0.02	NC	A	0.91	0	A	<0.01	NC	A	<0.01	NC	A
		<0.02			0.91			<0.01			<0.01		
Badging Facility	07-13-00	<0.02	NC	A	0.73	0	A	<0.01	NC	A	0.02	0	A
		<0.02			0.73			<0.01			0.02		
Area 2	07-18-00	<0.02	NC	A	1.1	0	A	<0.01	NC	A	<0.01	NC	A
		<0.02			1.1			<0.01			<0.01		
USGS 97	09-27-00	<0.041	NC	A	1.1	34	N	<0.006	NC	A	0.01 E	NC	N
		<0.041			1.8			<0.006			0.03		
USGS 82	09-28-00	<0.041	NC	A	0.56	1	A	<0.006	NC	A	0.02 E	NC	A
		<0.041			0.57			<0.006			0.02 E		
USGS 11	10-05-00	<0.041	NC	A	0.64	1	A	<0.006	NC	A	0.01 E	NC	A
		<0.041			0.64			<0.006			0.01 E		
USGS 124	10-05-00	<0.041	NC	A	0.77	0	A	<0.006	NC	A	<0.018	NC	A
		<0.041			0.78			<0.006			<0.018		
USGS 57	10-05-00	<0.041	NC	A	3.4	0	A	<0.006	NC	A	0.02	7	A
		<0.041			3.4			<0.006			0.02		
USGS 77	10-06-00	<0.041	NC	A	3.2	0	A	<0.006	NC	A	0.01 E	NC	A
		<0.041			3.2			<0.006			0.02 E		
USGS 50	10-10-00	<0.041	NC	A	43	1	A	<0.006	NC	A	0.04	0	A
		<0.041			44			<0.006			0.04		

Table 6. Measured concentrations, relative standard deviations, and acceptable or not acceptable reproducibility for ammonia, nitrate + nitrite, nitrite, and orthophosphate from replicate pairs collected from selected sites at the Idaho National Laboratory and vicinity, Idaho, 1996–2001.—Continued

[Locations of sites are shown in figures 1–4. Concentrations in table are rounded results, but relative standard deviations (RSDs) were calculated using unrounded results. **Reproducibility:** A, acceptable; N, not acceptable. **Abbreviations:** mg/L, milligrams per liter; N, nitrogen; P, phosphorous; E, estimated; <, less than; NC, not calculated because one or both concentrations were below the reporting level]

Site name	Sample collection date	Ammonia			Nitrate + nitrite			Nitrite			Orthophosphate		
		(mg/L as N)	RSD (percent)	Reproduc-ibility	(mg/L as N)	RSD (percent)	Reproduc-ibility	(mg/L as N)	RSD (percent)	Reproduc-ibility	(mg/L as P)	RSD (percent)	Reproduc-ibility
RWMC M13S	10-17-00	<0.041 <0.041	NC	A	0.34 0.35	1	A	<0.006 <0.006	NC	A	0.01 E 0.01 E	NC	A
USGS 104	10-23-00	<0.041 <0.041	NC	A	0.78 0.78	0	A	<0.006 <0.006	NC	A	0.01 E 0.01 E	NC	A
USGS 103	04-04-01	<0.041 <0.041	NC	A	0.76 0.77	0	A	<0.006 <0.006	NC	A	<0.018 0.01 E	NC	A
USGS 17	04-04-01	<0.041 <0.041	NC	A	0.38 0.38	0	A	<0.006 <0.006	NC	A	0.01 E 0.01 E	NC	A
Site 14	04-05-01	<0.041 <0.041	NC	A	0.55 0.56	1	A	<0.006 <0.006	NC	A	<0.018 <0.018	NC	A
USGS 9	04-11-01	<0.041 <0.041	NC	A	0.61 0.65	4	A	<0.006 <0.006	NC	A	0.01 E 0.01 E	NC	A
USGS 127	04-18-01	<0.041 <0.041	NC	A	0.55 0.56	0	A	<0.006 <0.006	NC	A	0.01 E 0.01 E	NC	A
CFA LF 2-10	04-30-01	<0.041 <0.041	NC	A	1.2 1.2	0	A	<0.006 <0.006	NC	A	0.045 0.01 E	NC	N
USGS 84	04-30-01	<0.041 <0.041	NC	A	0.74 0.74	0	A	<0.006 <0.006	NC	A	<0.018 0.011 E	NC	A
Site 9	07-05-01	<0.04 <0.04	NC	A	0.64 0.66	2	A	0.02 0.02	0	A	0.01 E 0.01 E	NC	A
USGS 29	07-16-01	<0.04 <0.04	NC	A	2.3 2.3	0	A	0.008 0.008	0	A	0.01 E 0.01 E	NC	A
Site 17	07-30-01	<0.04 <0.04	NC	A	1.1 1.1	0	A	0.004 E <0.006	NC	A	0.01 E 0.01 E	NC	A
USGS 38	10-11-01	<0.04 <0.04	NC	A	3.1 3.1	0	A	<0.008 <0.008	NC	A	0.02 0.02 E	NC	A
USGS 42	10-17-01	<0.04 <0.04	NC	A	1.6 1.6	1	A	<0.008 <0.008	NC	A	0.02 0.02	3	A
USGS 43	10-22-01	<0.04 0.02 E	NC	A	4.8 21	88	N	<0.008 <0.008	NC	A	0.02 E 0.03	NC	N

[1]Replicate pair concentrations were within ±0.5 times the reporting level.

Table 7. Measured concentrations, relative standard deviations, and acceptable or not acceptable reproducibility for selected metals from replicate pairs collected from selected sites at the Idaho National Laboratory and vicinity, Idaho, 1996–2001.

[Locations of sites are shown in figures 1–4. Concentrations in table are rounded results, but relative standard deviations (RSDs) were calculated using unrounded results. **Reproducibility:** A, acceptable; N, not acceptable. **Abbreviations:** μg/L, micrograms per liter; <, less than; E, estimated; NC, not calculated because one or both concentrations were below the reporting level; –, no data]

Site name	Sample collection date	Aluminum			Antimony			Arsenic			Barium		
		(μg/L)	RSD (percent)	Reproduc-ibility	(μg/L)	RSD (percent)	Reproduc-ibility	(μg/L)	RSD (percent)	Reproduc-ibility	(μg/L)	RSD (percent)	Reproduc-ibility
No Name 1	10-01-96	12	0	A	<1	NC	A	2	0	A	70	1	A
		12			<1			2			69		
ANP 9	10-27-97	4.3	2	A	<1	NC	A	2	28	N	86	1	A
		4.2			<1			3			85		
USGS 7	04-13-98	5	7	A	<1	NC	A	4	0	A	16	1	A
		4			<1			4			16		
USGS 65	04-28-98	5	2	A	<1	NC	A	1	0	A	49	0	A
		5			<1			1			49		
USGS 97	10-14-98	5	0	A	<1	NC	A	1	8	A	138	1	A
		5			<1			1			139		
PSTF Test	04-04-00	1	8	A	<1	NC	A	2	1	A	68	1	A
		1.3			<1			2			67		
USGS 98	04-04-00	3	43	N	<1	NC	A	1 E	NC	A	56	1	A
		5			<1			2 E			55		
USGS 26	04-12-00	2	0	A	<1	NC	A	2	7	A	36	0	A
		2			<1			3			36		
USGS 97	09-27-00	2	5	A	0.1	4	A	2	8	A	122	0	A
		2			0.1			2			123		
USGS 84	04-30-01	4	3	A	0.1	3	A	2 E	NC	A	81	0	A
		4			0.1			2 E			81		

Site name	Sample collection date	Beryllium			Cadmium			Cobalt			Copper		
		(μg/L)	RSD (percent)	Reproduc-ibility	(μg/L)	RSD (percent)	Reproduc-ibility	(μg/L)	RSD (percent)	Reproduc-ibility	(μg/L)	RSD (percent)	Reproduc-ibility
No Name 1	10-14-96	<1	NC	A	<1	NC	A	<1	NC	A	<1	NC	A
		<1			<1			<1			<1		
ANP 9	10-27-97	<1	NC	A	<1	NC	A	<1	NC	A	<1	NC	A
		<1			<1			<1			<1		
USGS 7	04-13-98	<1	NC	A	<1	NC	A	<1	NC	A	<1	NC	A
		<1			<1			<1			<1		
USGS 65	04-28-98	<1	NC	A	<1	NC	A	<1	NC	A	<1	NC	A
		<1			<1			<1			<1		
USGS 97	10-14-98	<1	NC	A	<1	NC	A	<1	NC	A	<1	NC	A
		<1			<1			<1			<1		
PSTF Test	04-04-00	<1	NC	A	<1	NC	A	<1	NC	A	<1	NC	A
		<1			<1			<1			<1		
USGS 98	04-04-00	<1	NC	A	<1	NC	A	<1	NC	A	1	4	A
		<1			<1			<1			1		
USGS 26	04-12-00	<1	NC	A	<1	NC	A	<1	NC	A	<1	NC	A
		<1			<1			<1			<1		
USGS 97	09-27-00	<0.06	NC	A	0.05	5	A	0.2	6	A	1	6	A
		<0.06			0.05			0.2			1		
USGS 84	04-30-01	<0.06	NC	A	0.2	0	A	0.08	3	A	2	1	A
		<0.06			0.2			0.09			2		

Table 7. Measured concentrations, relative standard deviations, and acceptable or not acceptable reproducibility for selected metals from replicate pairs collected from selected sites at the Idaho National Laboratory and vicinity, Idaho, 1996–2001.—Continued

[Locations of sites are shown in figures 1–4. Concentrations in table are rounded results, but relative standard deviations (RSDs) were calculated using unrounded results. **Reproducibility:** A, acceptable; N, not acceptable. **Abbreviations:** µg/L, micrograms per liter; <, less than; E, estimated; NC, not calculated because one or both concentrations were below the reporting level; –, no data]

Site name	Sample collection date	Lead (µg/L)	RSD (percent)	Reproduc-ibility	Manganese (µg/L)	RSD (percent)	Reproduc-ibility	Mercury (µg/L)	RSD (percent)	Reproduc-ibility	Molybdenum (µg/L)	RSD (percent)	Reproduc-ibility
No Name 1	10-14-96	<1 <1	NC	A	2 2	0	A	<0.1 <0.1	NC	A	6 6	0	A
ANP 9	10-27-97	<1 <1	NC	A	1 <1	NC	A	<0.1 <0.1	NC	A	2.9 2.9	1	A
USGS 7	04-13-98	<1 <1	NC	A	3 3	4	A	<0.1 <0.1	NC	A	4 4	1	A
USGS 65	04-28-98	2 ?	1	A	<1 <1	NC	A	<0.1 <0.1	NC	A	3 2	1	A
USGS 97	10-14-98	<1 <1	NC	A	<1 <1	NC	A	<0.1 <0.1	NC	A	1 1	5	A
PSTF Test	04-04-00	<1 <1	NC	A	<1 <1	NC	A	<0.23 <0.23	NC	A	2 2	1	A
USGS 98	04-04-00	3 3	0	A	<1 <1	NC	A	<0.23 <0.23	NC	A	<1 <1	NC	A
USGS 26	04-12-00	<1 <1	NC	A	<1 <1	NC	A	<0.23 <0.23	NC	A	3 3	0	A
USGS 97	09-27-00	0.8 0.8	1	A	<1 <1	NC	A	<0.23 <0.23	NC	A	1.5 1.5	1	A
USGS 84	04-30-01	12 12	0	A	0.2 0.3	26	N	<0.01 <0.01	NC	A	2 2	1	A

Site name	Sample collection date	Nickel (µg/L)	RSD (percent)	Reproduc-ibility	Selenium (µg/L)	RSD (percent)	Reproduc-ibility	Silver (µg/L)	RSD (percent)	Reproduc-ibility	Thallium (µg/L)	RSD (percent)	Reproduc-ibility
No Name 1	10-14-96	<1 <1	NC	A	– –	–	–	<1 <1	NC	A	<0.5 <0.5	NC	A
ANP 9	10-27-97	<1 <1	NC	A	– –	–	–	<1 <1	NC	A	<0.5 <0.5	NC	A
USGS 7	04-13-98	<1 <1	NC	A	– –	–	–	<1 <1	NC	A	<0.5 <0.5	NC	A
USGS 65	04-28-98	<1 <1	NC	A	2 2	13	A	<1 <1	NC	A	– –	–	–
USGS 97	10-14-98	<1 <1	NC	A	2 2	16	A[1]	<1 <1	NC	A	– –	–	–
PSTF Test	04-04-00	<1 <1	NC	A	– –	–	–	<1 <1	NC	A	<0.5 <0.5	NC	A
USGS 98	04-04-00	<1 <1	NC	A	<2.4 2.2 E	NC	A	<1 <1	NC	A	– –	–	–
USGS 26	04-12-00	<1 <1	NC	A	– –	–	–	<1 <1	NC	A	<0.5 <0.5	NC	A
USGS 97	09-27-00	2 1	22	N	2.0 E 1.8 E	NC	A	<1 <1	NC	A	– –	–	–
USGS 84	04-30-01	<0.06 <0.06	NC	A	1.7 E 1.8 E	NC	A	<1 <1	NC	A	– –	–	–

Table 7.   Measured concentrations, relative standard deviations, and acceptable or not acceptable reproducibility for selected metals from replicate pairs collected from selected sites at the Idaho National Laboratory and vicinity, Idaho, 1996–2001.—Continued

[Locations of sites are shown in figures 1–4. Concentrations in table are rounded results, but relative standard deviations (RSDs) were calculated using unrounded results. **Reproducibility:** A, acceptable; N, not acceptable. **Abbreviations:** μg/L, micrograms per liter; <, less than; E, estimated; NC, not calculated because one or both concentrations were below the reporting level; –, no data]

Site name	Sample collection date	Uranium			Zinc		
		(μg/L)	RSD (percent)	Reproduc-ibility	(μg/L)	RSD (percent)	Reproduc-ibility
No Name 1	10-14-96	2 2	0	A	3 2	28	N
ANP 9	10-27-97	2.3 2.3	1	A	12 11	6	A
USGS 7	04-13-98	2.3 2.3	1	A	2 1	40	N
USGS 65	04-28-98	2 2	1	A	368 366	0	A
USGS 97	10-14-98	2 2	0	A	91 90	0	A
PSTF Test	04-04-00	2 2	1	A	<1 1.5	NC	A
USGS 98	04-04-00	2 2	0	A	123 119	2	A
USGS 26	04-12-00	2 2	1	A	<1 <1	NC	A
USGS 97	09-27-00	2 2	0	A	108 102	4	A
USGS 84	04-30-01	2 2	0	A	391 390	0	A

[1]Replicate pair concentrations were within ±0.5 times the reporting level.

Table 8.   Measured concentrations, relative standard deviations, and acceptable or not acceptable reproducibility for total dissolved chromium and hexavalent chromium from replicate pairs collected from selected sites at the Idaho National Laboratory and vicinity, Idaho, 1996–2001.

[Locations of sites are shown in figures 1–4. Concentrations in table are rounded results, but relative standard deviations (RSDs) were calculated using unrounded results. **Reproducibility:** A, acceptable; N, not acceptable. **Abbreviations:** µg/L, micrograms per liter; <, less than; E, estimated; NC, not calculated because one or both concentrations were below the reporting level; –, no data]

Site name	Sample collection date	Total dissolved chromium			Hexavalent chromium		
		(µg/L)	RSD (percent)	Reproduc- ibility	(µg/L)	RSD (percent)	Reproduc- ibility
USGS 125	01-09-96	6 5	13	A	1 <1	NC	A
USGS 83	04-01-96	9 13	26	N	20 17	11	A
USGS 34	04-02-96	17 19	8	A	28 34	14	A
USGS 79	04-02-96	8 <5	NC	N	5 8	33	N
USGS 5	04-10-96	<5 <5	NC	A	1 2	47	N
USGS 9	04-16-96	5 5	0	A	2 2	0	A
MTR TEST	04-17-96	5 <5	NC	A	2 3	28	N
USGS 109	04-17-96	<5 5	NC	A	3 <1	NC	N
TRA DISP	04-18-96	13 14	5	A	5 3	35	N
Site 14	04-24-96	<5 7	NC	A	3 2	28	N
TRA 1	07-18-96	<5 <5	NC	A	<1 2	NC	N
TRA 3	07-18-96	<5 <5	NC	A	<1 <1	NC	A
USGS 66	07-25-96	35 40	9	A	24 27	8	A
USGS 72	07-30-96	<5 <5	NC	A	<1 1	NC	A
No Name 1	10-14-96	8 8	0	A	<1 <1	NC	A
USGS 27	10-15-96	9 8	8	A	<1 <1	NC	A
USGS 77	10-17-96	14 13	5	A	3 6	47	N
USGS 73	10-23-96	80 77	3	A	54 15	80	N
USGS 38	10-25-96	8 7	9	A	<1 <1	NC	A
USGS 54	01-23-97	16 15	5	A	– –	–	–

Table 8. Measured concentrations, relative standard deviations, and acceptable or not acceptable reproducibility for total dissolved chromium and hexavalent chromium from replicate pairs collected from selected sites at the Idaho National Laboratory and vicinity, Idaho, 1996–2001.—Continued

[Locations of sites are shown in figures 1–4. Concentrations in table are rounded results, but relative standard deviations (RSDs) were calculated using unrounded results. **Reproducibility:** A, acceptable; N, not acceptable. **Abbreviations:** µg/L, micrograms per liter; <, less than; E, estimated; NC, not calculated because one or both concentrations were below the reporting level; –, no data]

Site name	Sample collection date	Total dissolved chromium			Hexavalent chromium		
		(µg/L)	RSD (percent)	Reproduc- ibility	(µg/L)	RSD (percent)	Reproduc- ibility
USGS 60	04-02-97	10 10	3	A	– –	–	–
USGS 109	04-03-97	7 8	11	A	– –	–	–
USGS 86	04-03-97	13 11	13	A	– –	–	–
CPP 5	04-08-97	<5 15	NC	N	– –	–	–
NPR Test	04-08-97	7 8	7	A	– –	–	–
USGS 63	04-22-97	33 30	8	A	– –	–	–
USGS 101	04-28-97	<5 <5	NC	A	– –	–	–
USGS 58	05-05-97	14 14	1	A	– –	–	–
CWP 1	07-15-97	5 <5	NC	A	– –	–	–
USGS 108	10-07-97	11 10	5	A	– –	–	–
USGS 8	10-08-97	6 7	7	A	– –	–	–
PW 8	10-21-97	12 11	1	A	– –	–	–
SITE 4	10-21-97	7 7	8	A	– –	–	–
ANP 9	10-27-97	4 4	0	A	– –	–	–
USGS 5	03-31-98	<14 <14	NC	A	– –	–	–
USGS 19	04-08-98	<14 <14	NC	A	– –	–	–
USGS 7	04-13-98	4 4	7	A	– –	–	–
Highway 3	04-20-98	<14 <14	NC	A	– –	–	–
Arbor Test	04-22-98	<14 <14	NC	A	– –	–	–
USGS 65	04-28-98	174 174	0	A	– –	–	–

Table 8. Measured concentrations, relative standard deviations, and acceptable or not acceptable reproducibility for total dissolved chromium and hexavalent chromium from replicate pairs collected from selected sites at the Idaho National Laboratory and vicinity, Idaho, 1996–2001.—Continued

[Locations of sites are shown in figures 1–4. Concentrations in table are rounded results, but relative standard deviations (RSDs) were calculated using unrounded results. **Reproducibility:** A, acceptable; N, not acceptable. **Abbreviations:** μg/L, micrograms per liter; <, less than; E, estimated; NC, not calculated because one or both concentrations were below the reporting level; –, no data]

Site name	Sample collection date	Total dissolved chromium			Hexavalent chromium		
		(μg/L)	RSD (percent)	Reproduc-ibility	(μg/L)	RSD (percent)	Reproduc-ibility
USGS 69	07-01-98	<14	NC	A	–	–	–
		<14			–		
TRA 4	07-21-98	<14	NC	A	–	–	–
		<14			–		
CPP 2	09-30-98	<14	NC	A	–	–	–
		8 E			–		
USGS 97	10-14-98	8	3	A	–	–	–
		8			–		
USGS 1	10-20-98	<14	NC	A	–	–	–
		<14			–		
USGS 110A	10-20-98	<14	NC	A	–	–	–
		<14			–		
WS INEL 1	10-20-98	11 E	NC	A	–	–	–
		10 E			–		
USGS 68	12-01-98	<70	NC	A	–	–	–
		<70			–		
USGS 99	03-31-99	<14	NC	A	–	–	–
		<14			–		
USGS 79	04-01-99	<14	NC	A	–	–	–
		<14			–		
USGS 83	04-01-99	11 E	NC	A	–	–	–
		11 E			–		
USGS 12	04-05-99	<14	NC	A	–	–	–
		<14			–		
EBR 1	04-13-99	9 E	NC	A	–	–	–
		<14			–		
USGS 61	04-13-99	12 E	NC	A	–	–	–
		10 E			–		
P AND W 2	04-14-99	<14	NC	A	–	–	–
		<14			–		
USGS 100	04-19-99	<14	NC	A	–	–	–
		<14			–		
CPP 1	04-26-99	<14	NC	A	–	–	–
		<14			–		
CWP 3	06-30-99	<14	NC	A	–	–	–
		<14			–		
TRA 1	07-21-99	<14	NC	A	–	–	–
		<14			–		
USGS 11	10-07-99	15	NC	A	–	–	–
		7 E			–		

Table 8. Measured concentrations, relative standard deviations, and acceptable or not acceptable reproducibility for total dissolved chromium and hexavalent chromium from replicate pairs collected from selected sites at the Idaho National Laboratory and vicinity, Idaho, 1996–2001.—Continued

[Locations of sites are shown in figures 1–4. Concentrations in table are rounded results, but relative standard deviations (RSDs) were calculated using unrounded results. **Reproducibility:** A, acceptable; N, not acceptable. **Abbreviations:** μg/L, micrograms per liter; <, less than; E, estimated; NC, not calculated because one or both concentrations were below the reporting level; –, no data]

Site name	Sample collection date	Total dissolved chromium			Hexavalent chromium		
		(μg/L)	RSD (percent)	Reproduc-ibility	(μg/L)	RSD (percent)	Reproduc-ibility
USGS 34	10-14-99	9 E <14	NC	A	– –	–	–
USGS 58	10-21-99	16 12 E	NC	A	– –	–	–
USGS 105	10-25-99	10 E 8 E	NC	A	– –	–	–
USGS 23	10-27-99	<14 9 E	NC	A	– –	–	–
CFA 2	01-19-00	13 E 13 E	NC	A	– –	–	–
PSTF Test	04-04-00	2.9 2.8	4	A	– –	–	–
USGS 98	04-04-00	6.3 5.5	10	A	– –	–	–
USGS 12	04-05-00	<14 <14	NC	A	– –	–	–
MTR TEST	04-10-00	<14 <14	NC	A	– –	–	–
USGS 125	04-11-00	<14 <14	NC	A	– –	–	–
USGS 26	04-12-00	2.9 2.3	15	N	– –	–	–
USGS 27	04-12-00	<14 <14	NC	A	– –	–	–
CWP 2	07-17-00	<14 <14	NC	A	– –	–	–
USGS 66	07-25-00	<14 <14	NC	A	– –	–	–
USGS 62	09-27-00	9 E 8 E	NC	A	– –	–	–
USGS 97	09-27-00	6.1 6.1	0	A	– –	–	–
USGS 58	10-04-00	9 E 6 E	NC	A	– –	–	–
USGS 60	10-04-00	7 E <10	NC	A	– –	–	–
USGS 11	10-05-00	8 E <14	NC	A	– –	–	–

Table 8.   Measured concentrations, relative standard deviations, and acceptable or not acceptable reproducibility for total dissolved chromium and hexavalent chromium from replicate pairs collected from selected sites at the Idaho National Laboratory and vicinity, Idaho, 1996–2001.—Continued

[Locations of sites are shown in figures 1–4. Concentrations in table are rounded results, but relative standard deviations (RSDs) were calculated using unrounded results. **Reproducibility:** A, acceptable; N, not acceptable. **Abbreviations:** μg/L, micrograms per liter; <, less than; E, estimated; NC, not calculated because one or both concentrations were below the reporting level; –, no data]

Site name	Sample collection date	Total dissolved chromium			Hexavalent chromium		
		(μg/L)	RSD (percent)	Reproduc- ibility	(μg/L)	RSD (percent)	Reproduc- ibility
USGS 77	10-06-00	16 10 E	NC	A	– –	–	–
RWMC M13S	10-17-00	8 E 5 E	NC	A	– –	–	–
USGS 73	04-02-01	21 20	2	A	– –	–	–
USGS 103	04-04-01	8 E 7 E	NC	A	– –	–	–
USGS 17	04-04-01	<10 7 E	NC	A	– –	–	–
Site 14	04-05-01	8 E 6 E	NC	A	– –	–	–
USGS 9	04-11-01	8 E <10	NC	A	– –	–	–
RWMC M14S	04-17-01	17 16	5	A	– –	–	–
USGS 127	04-18-01	12 12	0	A	– –	–	–
CFA LF 2-10	04-30-01	13 10	13	A	– –	–	–
USGS 84	04-30-01	20 20	3	A	– –	–	–
USGS 126A	07-10-01	<10 <10	NC	A	– –	–	–
USGS 126B	07-10-01	<10 <10	NC	A	– –	–	–
TRA 3	07-19-01	5 E <10	NC	A	– –	–	–
Site 17	07-30-01	<10 <10	NC	A	– –	–	–
PW 8	10-09-01	6 E 7 E	NC	A	– –	–	–
USGS 38	10-11-01	<10 <10	NC	A	– –	–	–
Arbor Test	10-15-01	<10 <10	NC	A	– –	–	–

Table 9. Measured concentrations, relative standard deviations, and acceptable or not acceptable reproducibility for selected volatile organic compounds from replicate pairs collected from selected sites at the Idaho National Laboratory and vicinity, Idaho, 1996–2001.

[Locations of sites are shown in figures 1–4. Concentrations in table are rounded results, but relative standard deviations (RSDs) were calculated using unrounded concentration results. **Reproducibility:** A, acceptable; N, not acceptable. **Abbreviations:** µg/L, micrograms per liter; <, less than; E, estimated; NC, not calculated because one or both concentrations were below the reporting level; –, no data]

Site name	Sample collection date	1,1-dichloroethene µg/L	RSD (percent)	Reproducibility	Tetrachloroethene µg/L	RSD (percent)	Reproducibility	Tetrachloromethane µg/L	RSD (percent)	Reproducibility	Toluene µg/L	RSD (percent)	Reproducibility	1,1,1-Trichloroethane µg/L	RSD (percent)	Reproducibility	Trichloroethene µg/L	RSD (percent)	Reproducibility	Trichloromethane µg/L	RSD (percent)	Reproducibility
RWMC Production	01-16-96	<0.2	NC	A	<0.2	NC	A	4.4	0	A	<0.2	NC	A	0.6	0	A	2.3	3	A	0.7	0	A
		<0.2			<0.2			4.4			<0.2			0.6			2.4			0.7		
USGS 87	01-16-96	<0.2	NC	A	<0.2	NC	A	1.8	4	A	<0.2	NC	A	<0.2	NC	A	0.5	0	A	<0.2	NC	A
		<0.2			<0.2			1.7			<0.2			<0.2			0.5			<0.2		
USGS 34	04-02-96	<0.2	NC	A	<0.2	NC	A	<0.2	–	A	<0.2	NC	A	0.2	0	A	<0.2	NC	A	<0.2	NC	A
		<0.2			<0.2			<0.2			<0.2			0.2			<0.2			<0.2		
USGS 117	04-16-96	<0.2	NC	A	<0.2	NC	A	<0.2	–	A	<0.2	NC	A	<0.2	NC	A	<0.2	NC	A	<0.2	NC	A
		<0.2			<0.2			<0.2			<0.2			<0.2			<0.2			<0.2		
No Name 1	10-14-96	<0.2	NC	A	<0.2	NC	A	<0.2	–	A	<0.2	NC	A	<0.2	NC	A	<0.2	NC	A	<0.2	NC	A
		<0.2			<0.2			<0.2			<0.2			<0.2			<0.2			<0.2		
USGS 77	10-17-96	0.2	0	A	<0.2	NC	A	<0.2	–	A	<0.2	NC	A	0.4	0	A	<0.2	NC	A	<0.2	NC	A
		0.2			<0.2			<0.2			<0.2			0.4			<0.2			<0.2		
USGS 38	10-25-96	<0.2	NC	A	<0.2	NC	A	<0.2	–	A	<0.2	NC	A	0.3	28	N	<0.2	NC	A	<0.2	NC	A
		<0.2			<0.2			<0.2			<0.2			0.2			<0.2			<0.2		
USGS 107	10-01-97	<0.2	NC	A	<0.2	NC	A	<0.2	–	A	<0.2	NC	A	<0.2	NC	A	<0.2	NC	A	<0.2	NC	A
		<0.2			<0.2			<0.2			<0.2			<0.2			<0.2			<0.2		
ANP 9	10-27-97	<0.2	NC	A	<0.2	NC	A	<0.2	–	A	<0.2	NC	A	<0.2	NC	A	<0.2	NC	A	<0.2	NC	A
		<0.2			<0.2			<0.2			<0.2			<0.2			<0.2			<0.2		
USGS 117	04-01-98	<0.2	NC	A	<0.2	NC	A	<0.2	–	A	<0.2	NC	A	<0.2	NC	A	<0.2	NC	A	<0.2	NC	A
		<0.2			<0.2			<0.2			<0.2			<0.2			<0.2			<0.2		
USGS 7	04-13-98	<0.2	NC	A	<0.2	NC	A	<0.2	–	A	<0.2	NC	A	<0.2	NC	A	<0.2	NC	A	<0.2	NC	A
		<0.2			<0.2			<0.2			<0.2			<0.2			<0.2			<0.2		
USGS 65	04-28-98	<0.2	NC	A	<0.2	NC	A	<0.2	–	A	<0.2	NC	A	0.3	0	A	<0.2	NC	A	<0.2	NC	A
		<0.2			<0.2			<0.2			<0.2			0.3			<0.2			<0.2		
USGS 90	07-21-98	<0.2	NC	A	<0.2	NC	A	2.9	1	A	<0.2	NC	A	0.4	0	A	1.3	1	A	0.3	1	A
		<0.2			<0.2			2.9			<0.2			0.4			1.3			0.4		
USGS 97	10-14-98	<0.2	NC	A	<0.2	NC	A	<0.2	–	A	<0.2	NC	A	<0.2	NC	A	<0.2	NC	A	<0.2	NC	A
		<0.2			<0.2			<0.2			<0.2			<0.2			<0.2			<0.2		
USGS 88	01-13-99	<0.2	NC	A	0.4	39	N	1.6	2	A	0.4	30	N	0.2	1	A	0.7	1	A	0.5	0	A
		<0.2			0.2			1.6			0.3			0.2			0.7			0.5		
USGS 12	04-05-99	<0.2	NC	A	<0.2	NC	A	<0.2	–	A	<0.2	NC	A	<0.2	NC	A	<0.2	NC	A	<0.2	NC	A
		<0.2			<0.2			<0.2			<0.2			<0.2			<0.2			<0.2		

Table 9. Measured concentrations, relative standard deviations, and acceptable or not acceptable reproducibility for selected volatile organic compounds from replicate pairs collected from selected sites at the Idaho National Laboratory and vicinity, Idaho, 1996–2001.—Continued

[Locations of sites are shown in figures 1–4. Concentrations in table are rounded results, but relative standard deviations (RSDs) were calculated using unrounded concentration results. **Reproducibility:** A, acceptable; N, not acceptable. **Abbreviations:** µg/L, micrograms per liter; <, less than; E, estimated; NC, not calculated because one or both concentrations were below the reporting level; –, no data]

Site name	Sample collection date	1,1-dichloroethene (µg/L)	RSD (percent)	Reproducibility	Tetrachloroethene (µg/L)	RSD (percent)	Reproducibility	Tetrachloromethane (µg/L)	RSD (percent)	Reproducibility	Toluene (µg/L)	RSD (percent)	Reproducibility	1,1,1-Trichloroethane (µg/L)	RSD (percent)	Reproducibility	Trichloroethene (µg/L)	RSD (percent)	Reproducibility	Trichloromethane (µg/L)	RSD (percent)	Reproducibility
CPP 1	04-26-99	<0.2 / <0.2	NC	A	<0.2 / <0.2	NC	A	<0.2 / <0.2	–	A	<0.2 / <0.2	NC	A	<0.2 / <0.2	NC	A	<0.2 / <0.2	NC	A	<0.2 / <0.2	NC	A
RWMC M7S	07-12-99	<0.2 / <0.2	NC	A	0.3 / 0.3	2	A	5.5 / 5.5	0	A	<0.2 / <0.2	NC	A	0.7 / 0.7	0	A	2.2 / 2.2	0	A	0.5 / 0.6	2	A
USGS 34	10-14-99	<0.2 / <0.2	NC	A	<0.2 / <0.2	NC	A	<0.2 / <0.2	–	A	<0.2 / <0.2	NC	A	0.2 / 0.2	0	A	<0.2 / <0.2	NC	A	<0.2 / <0.2	NC	A
RWMC Production	01-13-00	<0.2 / <0.2	NC	A	0.26 / 0.26	1	A	5.0 / 5.0	1	A	<0.2 / <0.2	NC	A	0.6 / 0.6	1	A	2.5 / 2.5	1	A	0.9 / 0.9	1	A
PSTF Test	04-04-00	<0.2 / <0.2	NC	A	<0.2 / <0.2	NC	A	<0.2 / <0.2	–	A	<0.2 / <0.2	NC	A	<0.2 / <0.2	NC	A	<0.2 / <0.2	NC	A	<0.2 / <0.2	NC	A
USGS 98	04-04-00	<0.2 / <0.2	NC	A	<0.2 / <0.2	NC	A	<0.2 / <0.2	–	A	<0.2 / <0.2	NC	A	<0.2 / <0.2	NC	A	<0.2 / <0.2	NC	A	<0.2 / <0.2	NC	A
USGS 12	04-05-00	<0.2 / <0.2	NC	A	<0.2 / <0.2	NC	A	<0.2 / <0.2	–	A	<0.2 / <0.2	NC	A	<0.2 / <0.2	NC	A	<0.2 / <0.2	NC	A	<0.2 / <0.2	NC	A
USGS 119	04-06-00	<0.2 / <0.2	NC	A	<0.2 / <0.2	NC	A	0.3 / 0.2	3	A	<0.2 / <0.2	NC	A	<0.2 / <0.2	NC	A	<0.2 / <0.2	NC	A	<0.2 / <0.2	NC	A
USGS 26	04-12-00	<0.2 / <0.2	NC	A	<0.2 / <0.2	NC	A	<0.2 / <0.2	–	A	<0.2 / <0.2	NC	A	<0.2 / <0.2	NC	A	<0.2 / <0.2	NC	A	<0.2 / <0.2	NC	A
RWMC Production	07-13-00	<0.2 / <0.2	NC	A	0.2 / 0.2	5	A	4.9 / 5.1	3	A	<0.2 / <0.2	NC	A	0.5 / 0.5	2	A	2.3 / 2.4	2	A	0.9 / 0.9	3	A
USGS 97	09-27-00	<0.2 / <0.2	NC	A	<0.2 / <0.2	NC	A	<0.2 / <0.2	–	A	<0.2 / <0.2	NC	A	<0.2 / <0.2	NC	A	<0.2 / <0.2	NC	A	<0.2 / <0.2	NC	A
USGS 77	10-06-00	<0.2 / <0.2	NC	A	<0.2 / <0.2	NC	A	<0.2 / <0.2	–	A	<0.2 / <0.2	NC	A	0.2 / 0.2	3	A	<0.2 / <0.2	NC	A	<0.2 / <0.2	NC	A
USGS 88	01-23-01	<0.2 / <0.2	NC	A	<0.2 / <0.2	NC	A	1.1 / 1.1	2	A	<0.2 / <0.2	NC	A	0.1 / 0.1	2	A	0.5 / 0.5	2	A	0.4 / 0.4	1	A
USGS 84	04-30-01	<0.2 / <0.2	NC	A	<0.2 / <0.2	NC	A	<0.2 / <0.2	–	A	<0.2 / <0.2	NC	A	0.1 / 0.1	3	A	<0.2 / <0.2	NC	A	<0.2 / <0.2	NC	A
USGS 120	07-12-01	<0.2 / <0.2	NC	A	<0.2 / <0.2	NC	A	2.9 / 2.9	2	A	<0.2 / <0.2	NC	A	0.3 / 0.3	0	A	0.9 / 0.9	1	A	0.5 / 0.5	0	A
USGS 38	10-11-01	<0.2 / <0.2	NC	A	<0.2 / <0.2	NC	A	<0.2 / <0.2	NC	A	<0.2 / <0.2	NC	A	0.2 / 0.2	3	A	<0.2 / <0.2	NC	A	<0.2 / <0.2	NC	A

Table 10. Measured concentrations, relative standard deviations, and acceptable or not acceptable reproducibility for total organic carbon from replicate pairs collected from selected sites at the Idaho National Laboratory and vicinity, Idaho, 1996–2001.

[Locations of sites are shown in figures 1–4. Concentrations in table are rounded results, but relative standard deviations (RSDs) were calculated using unrounded results. **Reproducibility:** A, acceptable; N, not acceptable. **Abbreviations:** mg/L, milligrams per liter; E, estimated; NC, not calculated; <, less than; –, no data]

Site name	Sample collection date	Total organic carbon (mg/L)	RSD (percent)	Reproduc-ibility
NO NAME 1	10-14-96	0.6 1.3	52	N
USGS 27	10-15-96	0.7 0.7	0	A
USGS 77	10-17-96	0.5 0.3	35	N
USGS 38	10-25-96	0.5 0.8	33	N
USGS 107	10-01-97	0.3 0.3	0	A
USGS 108	10-07-97	<0.1 0.4	NC	N
USGS 8	10-08-97	0.5 0.1	94	N
USGS 101	10-16-97	1.0 0.2	94	N
ANP 9	10-27-97	<0.1 <0.1	NC	A
USGS 97	10-14-98	0.5 1	47	N
USGS 1	10-20-98	1.2 0.2	101	N
USGS 110A	10-20-98	1.5 2	20	N
USGS 11	10-07-99	<0.27 0.16 E	NC	A
USGS 34	10-14-99	0.23 E 0.41	NC	N
USGS 105	10-25-99	2.2 0.73	71	N
USGS 23	10-27-99	1.38 0.51	65	N
USGS 97	09-27-00	0.91 0.69	19	N
USGS 11	10-05-00	<0.27 <0.27	NC	A
USGS 77	10-06-00	0.15 E 0.17 E	NC	A
RWMC M13S	10-17-00	0.45 E 0.51 E	NC	A
USGS 38	10-11-01	0.34 E 0.39 E	NC	A

Table 11. Percentage of replicate pairs with concentrations with acceptable reproducibility for radiochemical, inorganic, and organic constituents collected from selected sites at the Idaho National Laboratory and vicinity, Idaho, 1996–2001.

[Values in **bold** type indicate that the percentage of replicate pairs with concentrations with acceptable reproducibility was less than 90 percent]

Constituent	Number of replicate pairs	Replicate pairs with concentrations with acceptable reproducibility Number	Replicate pairs with concentrations with acceptable reproducibility Percentage
Gross-alpha radioactivity	63	63	100
Gross-beta radioactivity	63	61	97
Cesium-137	93	91	98
Tritium	204	196	96
Strontium-90	123	119	97
Plutonium-238	28	28	100
Plutonium-239+240	28	28	100
Americium-241	28	27	96
Sodium	131	131	100
Chloride	202	197	98
Fluoride	7	7	100
Sulfate	83	83	100
Ammonia	98	91	93
Nitrate + nitrite	98	94	96
Nitrite	98	95	97
Orthophosphate	98	87	**89**
Aluminum	10	9	90
Antimony	10	10	100
Arsenic	10	9	90
Barium	10	10	100
Beryllium	10	10	100
Cadmium	10	10	100
Cobalt	10	10	100
Copper	10	10	100
Lead	10	10	100
Manganese	10	9	90
Mercury	10	10	100
Molybdenum	10	10	100
Nickel	10	9	90
Selenium	5	5	100
Silver	10	10	100
Thallium	5	5	100
Zinc	10	8	**80**
Chromium	97	93	96
Hexavalent chromium	19	10	**53**
Volatile organic compounds[1]	32	32	100
1,1-dichloroethene	32	32	100
Tetrachloroethene	32	31	97
Tetrachloromethane	32	32	100
Toluene	32	31	97
1,1,1-trichloroethane	32	31	97
Trichloroethene	32	32	100
Trichloromethane	32	32	100
Total organic carbon	21	8	**38**

[1]Includes all volatile organic compounds (VOCs) except for the seven VOCs listed in this table.

Table 12.   Ranges of concentrations, number of replicate pairs with calculated relative standard deviations, and pooled relative standard deviations for radiochemical, inorganic, and organic constituents, Idaho National Laboratory and vicinity, Idaho, 1996–2001.

[**Abbreviations:** RSD, relative standard deviation; pCi/L, picocuries per liter; Cs-137, cesium-137; mg/L, milligrams per liter; µg/L, micrograms per liter, N, nitrogen; P, phosphorous]

Constituent	Concentration range	Number of replicate pairs with calculated RSDs	Pooled RSD (percent)
Gross-beta radioactivity (pCi/L as Cs-137)	6.0–12	6	18
	50–60	1	17
Tritium (pCi/L)	500–<2,000	26	8.4
	2,000–<20,000	43	7.2
	20,000–80,000	5	1.3
Strontium-90 (pCi/L)	6.0–<12	8	23
	12–<25	10	6.1
	25–250	7	3.4
Sodium (mg/L)	5–<30	115	3.2
	30–180	15	2.6
	600–700	1	1.3
Chloride (mg/L)	3–<20	129	2.3
	20–<60	43	14
	60–350	30	[1]1.5
Fluoride (mg/L)	0.2–0.3	7	6.6
Sulfate (mg/L)	10–70	69	1.6
	100–400	13	1.5
	2,300–2,400	1	1.8
Ammonia (mg/L as N)	0.01–0.07	15	23
	0.7–0.8	1	0.9
Nitrate + Nitrite (mg/L as N)	0.3–< 0.7	30	2.0
	0.7–6.0	65	[1]3.9
	10–60	3	17
Nitrite (mg/L as N)	0.008–0.02	7	17
Orthophosphate (mg/L as P)	0.01–0.04	54	16
	0.2–0.3	1	7.6
Aluminum (µg/L)	1.0–12	10	12
Antimony (µg/L)	0.1–0.2	2	3.5
Arsenic (µg/L)	1.0–5.0	8	12
Barium (µg/L)	15–150	10	0.7
Cadmium (µg/L)	0.04–0.2	2	1.4
Cobalt (µg/L)	0.08–0.3	2	6.4
Copper (µg/L)	1.0–2.0	3	3.5
Lead (µg/L)	0.8–12	4	0.6
Manganese (µg/L)	0.2–3.0	3	4.4
Molybdenum (µg/L)	1.0–6.0	9	0.9
Nickel (µg/L)	1.0–2.0	1	22
Selenium (µg/L)	1.0–2.0	2	15
Uranium (µg/L)	1.0–3.0	10	0.7
Zinc (µg/L)	1.0–3.0	2	33
	10–400	6	1.2

Table 12.   Ranges of concentrations, number of replicate pairs with calculated relative standard deviations, and pooled relative standard deviations for radiochemical, inorganic, and organic constituents, Idaho National Laboratory and vicinity, Idaho, 1996–2001.—Continued

[**Abbreviations:** RSD, relative standard deviation; pCi/L, picocuries per liter; Cs-137, cesium-137; mg/L, milligrams per liter; μg/L, micrograms per liter; N, nitrogen; P, phosphorous]

Constituent	Concentration range	Number of replicate pairs with calculated RSDs	Pooled RSD (percent)
Chromium (μg/L)	2.0–<13	23	11
	13–40	13	5.1
	70–200	2	1.2
Hexavalent chromium (μg/L)	1.0–<10	7	40
	10–40	4	51
1,1-dichloroethene (μg/L)	0.1–0.3	1	0.0
Tetrachloroethene (μg/L)	0.2–0.4	4	21
Tetrachloromethane (μg/L)	0.2–6.0	10	1.8
Toluene (μg/L)	0.3–0.4	1	30
1,1,1-trichloroethane (μg/L)	0.1–0.7	16	5.6
Trichloroethene (μg/L)	0.4–2.5	9	2.1
Trichloromethane (μg/L)	0.3–1.0	8	2.0
Total organic carbon (mg/L)	0.3–1.8	13	60

[1]Two replicate pairs with switched sample bottles were excluded from calculation of the pooled RSD.

Table 13.   Measured concentrations of tritium, strontium-90, cesium-137, sodium, chloride, sulfate, chromium, hexavalent chromium, and ammonia from source-solution, field, and equipment blanks, Idaho National Laboratory and vicinity, Idaho, 1996–2001.

[Uncertainties for radionuclides are 1σ combined standard uncertainties. Values in **bold** indicate that contamination bias was considered present in the blank. **Abbreviations:** DIW, deionized water; IBW, certified inorganic-free blank water; pCi/L, picocuries per liter; mg/L, milligrams per liter; µg/L, micrograms per liter; N, nitrogen; E, estimated; <, less than; –, no data]

Blank collection date	Type of blank	Source solution	Tritium (pCi/L)	Strontium-90 (pCi/L)	Cesium-137 (pCi/L)	Sodium (mg/L)	Chloride (mg/L)	Sulfate (mg/L)	Chromium (µg/L)	Hexavalent chromium (µg/L)	Ammonia (mg/L as N)
10-28-96	Source	DIW	0±400	1.1±1.4	10±60	<0.20	<0.10	<0.10	<5	**2**	–
04-25-96	Field	DIW	-100±400	-0.2±1.5	2±22	–	**0.2**	–	<5	<1	–
10-28-96	Field	DIW	150±460	-0.3±1.4	–	<0.20	<0.10	<0.10	–	–	–
01-23-97	Field	DIW	60±220	1.3±1.6	–	–	<0.10	–	–	–	–
07-22-97	Field	DIW	30±220	0.3±1.4	–	–	<0.10	–	–	–	–
10-30-97	Field	DIW	140±220	0.7±1.6	–	<0.20	<0.10	<0.10	–	–	–
01-22-98	Field	DIW	10±240	0.8±1	–	–	–	–	–	–	–
04-13-98	Field	DIW	-40±220	0.5±1.4	-10±80	–	<0.10	–	–	–	–
10-27-98	Field	DIW	90±220	0.6±1.2		<0.06	<0.10	<0.10	<14	–	–
01-14-99	Field	IBW	-90±240	-0.7±1.2	–	–	<0.10	–	–	–	–
04-15-99	Field	IBW	-110±220	0.2±1.6	30±60	–	<0.10	–	–	–	–
04-20-99	Field[1]	IBW	0±220	0.7±1.4	13±40	<0.06	<0.10	<0.10	<1	–	**0.03**
07-08-99	Field	IBW	-40±220	-0.2±1.6		<0.06	<0.10	<0.10	<14	–	–
04-24-00	Field	DIW	-190±220	1.4±1.6	-20±80	–	<0.29	–	–	–	–
10-10-01	Field	IBW	-40±240	0.1±1.3		<0.09	<0.33	E0.05	<10	–	–
04-25-96	Equipment	DIW	0±400	-0.8±1.4	-50±80	–	<0.10	–	<5	<1	–
07-25-96	Equipment	DIW	-100±400	0.5±1.4		**0.5**	**0.2**	<0.10	<5	<1	–
10-28-96	Equipment	DIW	-400±400	-0.7±1.4	-50±60	<0.20	<0.10	<0.10	<5	**2**	–
01-23-97	Equipment	DIW	70±220	-0.7±1.6	–	–	**0.1**	–	–	–	–
07-22-97	Equipment	DIW	10±220	0.2±1.4	–	–	<0.10	–	–	–	–
10-29-97	Equipment	DIW	50±220	2±1.6	–	<0.20	<0.10	<0.10	<5	–	–
10-30-97	Equipment	DIW	10±220	-0.3±1.6	–	<0.20	<0.10	<0.10	–	–	–
04-13-98	Equipment	DIW	-10±220	0.5±1.4	10±40	–	**0.55**	–	–	–	–
10-27-98	Equipment	DIW	160±240	0.6±1	–	**0.73**	<0.10	<0.10	<14	–	–
07-08-99	Equipment	IBW	-10±220	-0.1±1.4	–	**0.13**	**0.15**	**0.57**	<14	–	–
04-24-00	Equipment	DIW	-140±220	1.3±1.6	-20±80	–	E 0.16	–	–	–	–
10-10-01	Equipment	IBW	0±240	-1.1±1.4	–	<0.09	<0.33	<0.10	<10	–	–

[1]The field blank on April 20, 1999, also included results for gross-alpha, gross-beta, americium-241, plutonium-238, and plutonium-239 +240 radioactivity and nitrate + nitrite, nitrite, orthophosphate, metals, and volatile organic compound concentrations. Results for all of these constituents were less than the minimum detectable concentration or reporting level.

Table 14. Probability of success, confidence level, maximum concentration in blank samples, and minimum concentration in environmental samples, Idaho National Laboratory and vicinity, Idaho, 1996–2001.

[The minimum concentration in environmental samples was the minimum concentration from all sampling sites (field blank) or from the 21 wells where portable sampling equipment was used to collect the water samples (equipment blanks). **Probability of Success:** Probability that each blank (or environmental) sample is less than the $m+1$ sample concentration. **Abbreviation:** mg/L, milligrams per liter]

Constituent	Blank type	Number of blank results (n)	Probability of success (p)	Confidence level (cl) (percent)	Maximum concentration in blank samples ($m+1$) (mg/L)	Minimum concentration in environmental samples (mg/L)
Chloride	Field	13	0.8	95	0.2	2.6
	Equipment	12	0.8	93	0.55	3.1
Sodium	Equipment	7	0.7	91	0.73	6.1
Sulfate	Equipment	7	0.7	91	0.57	13

Glossary

Bias A persistent positive or negative deviation of the mean value, obtained by using a specific method or procedure, from the true value.

Blank samples Used to identify potential sources of sample contamination and assess the magnitude of contamination with respect to target analytes (U.S. Geological Survey, 2006). Blanks discussed in this report include source-solution, equipment, and field blanks. Equipment blanks were collected at a sampling site and processed through portable sampling equipment (either a bailer or portable pump and tubing). Field blanks were collected at a sampling site and collected exactly like equipment blanks except that they were not processed through portable sampling equipment.

Blind sample A sample submitted for laboratory analysis with composition known to the submitter but unknown to the analyst (U.S. Geological Survey, 2006)

Combined standard uncertainty Standard uncertainty estimate reported at the 1σ confidence level by combining the standard uncertainties of the analysis (McCurdy and others, 2008, p. 18). For analyses performed at the U.S. Department of Energy Radiological Environmental Services Laboratory, uncertainties may include yields, half-lives, counting efficiencies, and counting times (Williams, 1997, p. 10).

Quality assurance A term used to describe programs and the sets of procedures, including (but not limited to) quality control procedures, which are necessary to assure data reliability (Friedman and Erdmann, 1982).

Quality control A term used to describe the routine procedures used to regulate measurements and produce data of satisfactory quality (Friedman and Erdmann, 1982).

Quality-control data Data from blank, replicate, reference, or spike samples. The data are used "to identify, quantify, and document bias and variability in data resulting from the collection, processing, shipping, and handling of samples" from field and laboratory personnel (U.S. Geological Survey, 2006, p. 133).

Quality-control samples Blank, replicate, reference, or spike samples.

Reliability "A statement of the error or precision of an estimate" (Spiegel, 1998, p. 194).

Replicate samples "Replicates—environmental samples collected in duplicate, triplicate, or greater multiples—are considered identical or nearly identical in composition and are analyzed for the same chemical properties" (U.S. Geological Survey, 2006, p. 143).

Reproducability "The closeness of agreement between individual results" (Kateman and Buydens, 1993, p. 11).

Variability "The degree of random error in independent measurements of the same quantity" (Mueller, 1998, p. vii).

www.ingramcontent.com/pod-product-compliance
Lightning Source LLC
Chambersburg PA
CBHW081559170526
45166CB00009B/2754